Poisoned Power

Poisoned Power

The Case Against Nuclear Power Plants Before and After Three Mile Island

by
John W. Gofman, Ph.D., M.D.
and Arthur R. Tamplin, Ph.D.

 **Rodale Press,** Emmaus, Pa.

Printed in the United States of America on recycled paper containing a
high percentage of de-inked fiber.

Library of Congress Cataloging in Publication Data
Gofman, John William.
   Poisoned power.
   Includes bibliographical references and index.
   1. Atomic power-plants—Environmental aspects.
2. Radioactive pollution.   I. Tamplin, Arthur R.,
joint author.   II. Title.
TD195.E4G63   1979   333.7   79-16781
ISBN 0-87857-288-0

  4 6 8 10 9 7 5 3

TABLE OF CONTENTS

Foreword to the 1979 Printing of Poisoned Power, After the Three Mile Island Near-Disaster

Both of us who wrote *Poisoned Power* find ourselves heartsick that it took a near-disaster for the American people to learn all the things which we described in this book eight years ago. There is nothing in *Poisoned Power* which is out-of-date, except for the names of a few characters in the shabby nuclear energy charade.

Huge quantities of hydrogen were generated in the recent Three Mile Island nuclear plant accident. A hydrogen explosion did indeed occur. It is largely a matter of luck that the hydrogen explosion was not much larger, considering the amount of hydrogen generated. Under slightly altered circumstances, the explosion could have ruptured the containment vessel or have led to a far more severe disruption of the reactor itself, in either case with possible release of massive quantities of radioactive poisons.

Yet, ever since the accident, the nuclear industry propagandists have been congratulating themselves on the fact that it was just a *near*-disaster and that tens or hundreds of thousands of people were not killed. But an actual disaster was prevented by a simple quirk of fate—when and how the hydrogen explosion occurred—and *not* by their engineering skills nor by their "defense in depth."

It is now clear to most Americans that the nuclear emperor is wearing no clothes. Those Americans who did read *Poisoned Power* knew this, and much more, long ago. While it is sad that we have waited so long to rid America of this monstrous aberration of technology, at least there is a better chance of doing so now, before many more Americans of this and future generations are condemned to miserable deaths.

This book is about health—the health of this and future generations. It is about the lies, the cover-ups, and the callousness of those who are willing to trick you into accepting nuclear power so that they (or their bosses) can make money or expand a bureaucratic empire, even though their activity kills people.

Nuclear power is not the only enterprise imposing on your health, but if allowed to proceed unchecked, it surely will lead ultimately to setting back public health by hundreds of years. And the public health disaster will then be irreversible because once the radioactive poisons are let loose into the environment, there is no way of bringing them back under control.

The Three Mile Island accident has left tremendous numbers of Americans interested in, and anxious to know about, how radiation affects health. This book provides a simple and clear explanation. In addition, it exposes the moral corruption of scientists, lawyers, physicians, industrialists, and government leaders in attempting to deceive the public into believing that there exists such a thing as a "safe," "permissible," or "allowable" dose of radiation which will do no harm.

This book tells the reader why cancer and leukemia risks are increased even with the smallest doses of radiation. The so-called permissible dose of radiation, for nuclear workers or for the public at large, represents only a legalized permit for the nuclear industry to commit random, premeditated murders upon the American population.

Following the crimes of the Nazis in the early 1940s, the Nuremberg Trials were held. There, it was declared for all the world to know that individuals must be held personally responsible for crimes against humanity and that it is not acceptable to say, "I was only following orders" or "I was only following government laws and policies."

Experimentation on humans without their knowledge or consent is obviously a crime. Taking life without due process of law is obviously a crime. There can be no doubt that the promoters of nuclear power—be they engineers, politicians, or scientists—are indeed committing these crimes against humanity. Americans would be justified in demanding that Nuremberg-type trials be held for these individuals.

Violating the Inalienable Right to Life

The charge that nuclear power promotion represents a crime against humanity is a serious one indeed. We do not make this charge lightly. So we must explain further.

After Three Mile Island, all Americans witnessed the

efforts of government and industry officials to say that the amount of radiation inflicted on people in the area caused no injuries or deaths. This is an absolute lie, as you will understand after reading this book. You will realize that we do indeed know the effects of so-called ''low'' doses of radiation.

However, we shall never know *exactly* how many people will die of premature cancer and leukemia from the Three Mile accident because in the critical first few days there were no radiation monitors in many of the areas where people live. The few monitors right around the plant and the few measurements from airplanes represent a travesty upon honest monitoring.

Believe it or not, the Nuclear Regulatory Commission (NRC) permits giant nuclear plants to operate *without* having monitors installed and operating at all times in every populated area within 50 miles of the plants. Consequently, during an accident, there can be no way for people to evaluate the danger they are in, no way for them to decide whether or not to stay. But the no-monitor policy is no accident, for it helps to protect the utilities from lawsuits for personal injury.

The number of premature deaths caused by the Three Mile Island accident *will be no fewer than six*. The number could easily be 60, or 600, for the doses could well be 100 times higher than the government estimates. And all the while, the industry and government officials lie to the American people, saying there was ''no injury to people.''

There are two possible ways to describe the motives of the promoters of nuclear power, yet either way makes them indictable for crimes against humanity.

First, let us assume that they really are ignorant about existing knowledge of the effects of ''low'' doses of radiation

X

when they say, "We don't really know yet about the effects of 'low' doses of radiation." In that case, these promoters of nuclear power are saying in effect, "Expose people first; learn the effects later." There is only one description for such planned mass experimentation on humans—moral depravity. And such experimentation with "low" doses of radiation can produce irreversible effects not only on this generation, but upon countless future generations of humans who have no voice, no choice. If that is not a crime against humanity, what is?

Alternatively, let us assume that they truly do know the facts about fatal injury from "low" doses of radiation, and yet they are still willing to promote nuclear power. In this case, the charge is not experimentation upon humans, but rather it is planned, random murder. The crime of murder is perhaps worse than the crime of experimentation.

And it is not only "low" doses of radiation, with their delayed effects, which are at issue. We could have had high, massive doses of radiation affecting hundreds of thousands of humans from the Three Mile Island accident, and that would have meant massive numbers of deaths, not later from cancer, but sooner from the agony of acute radiation sickness.

Yet the President of the United States, Jimmy Carter, urges that we go ahead with nuclear power even faster than before. He urges this with his accelerated licensing bill even as he is appointing a committee to figure out why the accident occurred. Obviously, his philosophy is also experimentation on humans. "Expose first; learn later."

The clamor of the nuclear industry to go full speed ahead, the clamor of the Carter-and-Schlesinger mentalities to go full speed ahead with nuclear power since we happened to escape a

XI

full-scale disaster at Three Mile Island, should teach the American people that nuclear power promoters are intent on "going for the Big Apple" before they will quit. Too much money is at stake. Nothing short of the loss of a major city—a New York, or Philadelphia, or Chicago, or Boston—will make them feel remorse.

But they can be stopped, with a very simple campaign.

Revolt of the Guinea Pigs

The pollsters are reporting that even after Three Mile Island, most people still believe we need nuclear power for the economy and still favor going ahead with it—provided the plants are near someone else. The willingness to risk killing other people in order to get what you think you need (nuclear power) is, of course, morally bankrupt and no different from the position of the active promoters.

Nevertheless, we are sure the pollsters would get a different answer if they asked the proper questions. Suppose they were to ask people:

"If you had proof that neither the government nor the nuclear power industry believes in the safety of nuclear plants, would you still favor going ahead with nuclear power?"

The evidence that neither government nor industry believes in the safety of nukes is the Price-Anderson Act, a law which is explained in this book. Utility officials told Congress that if they had to take financial responsibility for the death and havoc which their nukes might create, they would not build a single one. So Congress agreed that utility stockholders should not have to gamble on the safety of nukes, and in 1957,

XII

Congress passed the Price-Anderson Act which says the public will pay for the havoc. In 1967, when the utilities said they would not go ahead unless the Act were renewed, Congress renewed it. And in 1975, when the utilities said they could not afford to go ahead unless the law was renewed again, guess what! Congress renewed it for them.

The evidence is clear: utilities and Congress have no confidence in the safety of nukes. Utilities are willing to risk *your* lives and *your* property on nuclear power, but they are not willing to risk *their* dollars.

And so the most dangerous industry ever conceived is permitted by Congress to operate without even the normal restraint on reckless activity, namely the financial responsibility for the injury and damage caused by such activity. Obviously Congress has always thought that this is the treatment the American people deserve.

People who are sick about being used as guinea pigs had better start the 1980 campaign now. We urge 100 million Americans to tell their senators and representatives, "You sponsor repeal of the Price-Anderson Act now, or we are starting the campaign against you right now."

When the Price-Anderson Act is repealed and the utility companies, their suppliers, like Westinghouse and General Electric, and the banks have their own dollars gambled on the line for nuclear power, we will soon see how much confidence those in the nuclear power industry truly have in the safety of nuclear power.

We predict this simple campaign can persuade the industry to "phase out" nuclear power about 24 hours after repeal of that law.

XIII

"But we need the power . . ."

Lastly, we must deal with the economic blackmail being used to bludgeon Americans into accepting nuclear power. "It is either nuclear power, or starving in the dark," say the promoters of nuclear power. They lie like carpets.

We can say, with great assurance, that *nuclear power itself* is the greatest threat to our energy supply, to the health of our economy, and to employment for our people. If we stopped pouring funds down the rathole of nuclear power, the money would be available to stimulate bigger and cheaper sources of clean energy which are benign and which would be a real boon to our economy. We now have the equivalent of 50 giant nukes which are operable (sometimes) in this country. This book tells you about two simple, proven sources of additional energy, ready to "go" quickly, which would be equivalent to over 600 giant nukes (see page 206).

If there is one thing we do not have to worry about, it is that banning nuclear power will hurt either our energy supply or our economy. Nuclear power today provides about 12 percent of our electricity, but since electricity accounts for only a fraction of our energy, nuclear power is really supplying only 3½ percent of our total energy today. It is trivial.

Nuclear power is not necessary, it never has been necessary, and it never will be necessary.

How to Make Some Sense of "Millirems" and "Picocuries"

Virtually all Americans have heard by now of the "millirem" in news reports about the radiation doses received by people near the Three Mile Island plant. On page 44 of this book, it is pointed out that, for the kinds of radiation inflicted by that accident, the millirem is the same as the millirad—the unit used in this book.

It is a simple calculation indeed to estimate the number of fatal cancers and leukemias which are produced by any nuclear accident, provided we know the dose received. The process, which is explained in the following paragraphs, can be summarized in a glance by a single equation:

$$\frac{(\# \text{ of rems})}{\text{hour}} \times (\# \text{ of hours}) \times \frac{(\# \text{ of persons exposed})}{300 \text{ person-rems}} \times (1 \text{ death}) = \begin{array}{l} \# \text{ of deaths} \\ \text{which will} \\ \text{occur later} \end{array}$$

The first term or factor is read "number of rems per hour," and the fourth term is read "one death per 300 person-rems." Since the persons, rems, and hours in the numerator and denominators all cancel each other out, you are left with the number of deaths.

At Three Mile Island, the monumental nonpreparation of the Nuclear Regulatory Commission and the utility company left us without measurement of the true doses received by a million or so people. A few days after the accident began, the NRC began reporting radiation doses to the public as "a few millirems per hour." Invariably it was pointed out that the doses were "low," while such reports carefully avoided stating that it is not just the millirems per hour which matters, but rather the *total* number of millirems received by the people. And the *total* millirems are obtained by multiplying the millirems per hour by the number of hours of exposure, as shown by the first two terms in the equation. To convert millirems to rems, you just divide by 1,000 since there are 1,000 millirems per rem.

XV

Everyone knows that exposure continued at least 100 hours, possibly two or three times that. But the publicists for the nuclear industry realized that total millirems would be a much larger number than millirems per hour, so they helped to deceive the public about the true magnitude of the hazard by sticking to the number which sounded smaller.

Let us say, for illustrative purposes, that the average dose-rate during the first five days was only one millirem per hour, or 0.001 rem per hour. This is what we put into our first term of the equation.

Then let us say that the dose continued for 100 hours. So that's what we put in the second term of the equation. When we multiply terms one and two, the hours cancel out, and we get the total number of rems: 0.1 rem, or 100 millirems.

Since 950,000 people live in the four counties nearest the Three Mile Island plant, we can put in 1,000,000 as the third term in the equation. When we multiply 0.1 rem times 1,000,000 persons, we get 100,000 person-rems or man-rems.

Then we come to filling in the fourth term. In the first printing of *Poisoned Power,* we used figures for this term which were called *over*-estimates of the hazard by atomic energy "experts." Those "experts" were wrong then, and they are even more wrong now, for new evidence in the last eight years shows that we had *under*estimated the hazard. It takes *fewer* man-rems of radiation exposure to guarantee that someone will get cancer or leukemia than we realized eight years ago. Our best estimate is that there will be one death for every 300 man-rems received by a population whose *average* age at exposure is about 25 years. If we use the figure of 300 man-rems, we may *still* be underestimating the hazard. By the way, there will be one cancer whether 10 rems is given to each of 30 people (10 × 30 = 300), or whether 1 rem is given to each of 300 people (1 × 300 = 300), or whether 0.1 rem is given to 3,000 people (0.1 × 3,000 = 300), or whether 0.01 rem (10 millirems) is given to 30,000 people (0.01 × 30,000 = 300), because the number of man-rems is the same in each case; and that means the number of cancers which will occur later is the same.

Now we are ready to solve our equation. We have 100,000 man-rems to multiply by one death per 300 man-rems. The man-rems cancel out, and by simple division of 100,000 by 300, we arrive at 333 fatal cancers or leukemias. So it should now be clear that even very "low" dose-rates like one millirem per hour can kill plenty of people when the dose is inflicted for a few days on a large number of people. The true dose at Three Mile Island itself and the true number of cancers will never be known.

In view of these facts, it was clearly a deception for officials to talk about doses that are "only like getting an X-ray" as if giving a million people a medical X-ray were safe. Medical irradiation not only kills about 20,000 to 40,000 Americans per year, but medical irradiation is voluntary. There is a world of difference between risks which are freely chosen (like medical X-rays, smoking, driving) and deaths which are inflicted at random on people whose consent was never given—which certainly includes people of future generations who will be killed by the nuclear pollution we are creating today.

As they look at the figures used in our fourth term (one death per 300 man-rems), no doubt some so-called atomic

XVII

energy "experts" will again say we are overestimating the hazard of radiation. We are proud that we disagree with the "think-alikes" who populate the advisory committees and high government posts in public health. Their jobs, grants, and appointments depend on their minimizing the harmfulness of radiation so that the nuclear industry can go forward as their bosses desire.

Our purpose is the objective analysis of the facts at our disposal. Small wonder there is disagreement! With our degrees, experience, credentials, and proven abilities, we could each be earning $1,000 per day as consultants *for* the nuclear power industry, if only we were willing to confuse the public, muddy the logic, ignore enough evidence, and say that nuclear power plants make "good neighbors." But we won't take the industry's blood money. You decide whose credibility is higher, when experts disagree.

Curies, Microcuries, and Picocuries

Now we come to the term "curie," a radiation measurement we avoided in the first printing of *Poisoned Power* simply because it is quite complex to go from the curie measure to the rad or rem measure, and rads or rems are what are really important for health effects.

However, the Three Mile Island accident has introduced the term "picocurie" into news reports about levels of radioactivity in milk, for instance, "picocuries per liter of milk." A liter is about a quart, but what is a picocurie?

In nuclear science, we often need a description of how many radioactive atoms are disintegrating by radioactive decay in one second. Disintegrations are emissions by the atoms of alpha, beta, or gamma rays with various energies. So a curie is a measure of the "strength" of a radioactive source.

By universal scientific agreement, we state that when we have a radioactive substance which is showing 37 billion disintegrations per second, we have one curie of radioactivity. It does not matter if we are talking about plutonium, strontium, iodine, or any other radioactive species or mixture of species; it is always 37 billion disintegrations per second which is called one curie.

In order to describe radioactive sources which are less strong than a curie, we have sub-units:

> *one millicurie* means the substance makes 37 million emissions per second;
> *one microcurie* means it makes 37 thousand emissions per second;
> *one nanocurie* means it makes 37 emissions per second;
> *one picocurie* means it makes 0.037 emissions every second.

From Curies and Picocuries to Rems and Millirems

When the public hears that 100 millicuries of radioactive iodine have been released from a nuclear plant, the immediate question is, "How big a dose (in millirems) will that give me?"

The answer is far from simple to figure out, for readily understood reasons.

In order for any of those "millicuries" to produce "millirems" in people, the radioactive material must first reach people.

First of all, the direction of the wind and its speed determine how many millicuries will reach a particular region in a certain amount of time. Some radioactive substances will decay in seconds and never reach people; others will be around for centuries or longer.

Next, there is the question of whether the radiation exposure from the millicuries is only outside the body, or whether some of the radioactive material gets *into* the body, via inhalation or drinking and eating contaminated water and food. If contaminated food is shipped and sold to people at a distance from the release of radioactivity, they can become irradiated by a "low" level release even though they are a hundred miles away.

For instance, radioactive substances can deposit themselves on the ground and fields, and cows foraging the area like vacuum cleaners can take up the radioactive substances into their bodies. Later, the cows can secrete radioactivity into their milk. If milk from a particular dairy is sold 100 miles away, the child who drinks the milk will get exposed to radiation even though his parents might assume they were a safe distance from the radioactive release.

There is a large body of scientific investigation devoted to finding out the likelihood of deposition of radioactive poisons at various distances from a nuclear plant after a release. There are still many uncertainties in the data for such deposition.

In addition, there is a large body of investigation concerning how much of what is deposited on land is likely to get into cow's milk or into the flesh of animals which are used for food.

And in addition to that research, there is a large body of

investigation concerning how much of whatever radioactive material is eaten will stay in the body, and for how long.

Lastly, calculations have to be made of the radiation dose (in millirems) caused by each different kind of radioactive atom which does stay in the body. Different radioactive substances give off different amounts of energy with each radioactive disintegration, and it is this energy which is causing the injury to the cell. The Nuclear Regulatory Commission now publishes tables which provide estimates of the number of millirems harmful to various organs of the human body which will be given off by every picocurie of a variety of radioactive substances actually eaten.

Every one of the factors just described is involved in trying to convert picocuries released at a nuclear facility into millirems of radiation dose received by the public. However, the dose-estimates reported around the Three Mile Island accident did *not* take all these factors into account. When measurements were finally made there, they were direct readings of only the external gamma ray dose coming primarily from radioactive gases in the air.

A recent study by some independent scientists at Heidelberg University is very critical of estimates made by the nuclear industry and by government about what happens to the picocuries at every step of the chain, from release right through to the ultimate estimate of millirems inflicted on people. The Heidelberg study suggests that the nuclear industry and government always choose those numbers which will make it appear that the dose to people is lower than it truly is, with the dose sometimes understated by as much as 1,000-fold.

XXI

Since the whole field (converting picocuries to millirems) is complicated, it is hard to know sometimes where the truth lies in such matters. One point is certain, however:

An ethical society, concerned with preserving the inalienable right to life, would learn all the steps in such pathways before ever permitting activities which could release the radioactive poisons upon the public. An uncertainty factor of 1,000 is a horrible uncertainty to have about the dose a human infant will receive. Experimentation on people by the nuclear industry must be stopped, and the industry's disdain for people's health—its "Expose first; learn later" philosophy—must be exposed for its moral bankruptcy.

The risks from irradiation are cumulative. A small dose will give you a small risk. But *another* small dose will give you an *addititional* small risk. By now, the nuclear industry must have announced 100,000 "small" releases of radioactivity to the environment. It is the only industry which can add 100,000 "small" releases to each other, and still say the sum is small and the harm to the public is zero!

Summary: The Important Questions

There has been much press and TV coverage devoted to the technical aspects of the Three Mile Island accident, but very little to its moral aspects. Yet the really important questions about nuclear power are ethical:

—The use of lies and deception by the nuclear
  industry in order to manipulate public opinion, and
  in order to *use* people, even kill people, for the
  benefit of that industry.

XXII

—The experimentation on people without their
knowledge or consent.
—The acceptance of random murder and denial of
the inalienable right to life as the cost of
''progress.''
—The genetic degradation of the human species, vs.
our minimum responsibility to protect the species'
genes from injury.
—The need to hold bureaucrats and industry
employees personally accountable and responsible
for implementing hazardous and even murderous
policies, even if such policies are advocated by
Congress and the President.

Yes, *Poisoned Power* is a sad story about the absence of
ethics and morals in men. But it is not too late to jolt society
into realization of what is going on, and what is in the future if
humans do not improve in the very basic and minimum
principles of morality. Either we improve, or the future is
dismal indeed. We hope that *Poisoned Power* upsets you
enough to make you work toward such improvement.

—John W. Gofman
San Francisco
June 1979

Foreword

I don't happen to like the title of this book. As for the book itself, I hope that millions of people will read it, because nuclear pollution is certainly a most serious threat to life.

Exposure to nuclear radiation can cause cancer, it can cause babies to be born mentally or physically defective, and it can cause increases in many serious illnesses like heart disease. I know of no one who denies these statements.

Fortunately, there is a chance to prevent serious nuclear pollution; it has not yet occurred. The threat, however, lies in the country's growing commitment to nuclear power plants for electricity, and to nuclear weapons for defense.

The problem with nuclear electricity is that as much long-lived radioactivity is produced inside one large nuclear power plant every year as there is in the explosion of about 1,000 Hiroshima bombs.

When we say "long-lived" radioactivity, we mean

long. Some kinds last for 100, 300, and even 240,000 years before decaying fully.

Unprotected, above-ground nuclear power plants, loaded with radioactivity in their cores, would certainly be large liabilities if this country were ever under attack. They seem to make the country virtually indefensible.

Quite aside from war or sabotage, an accident allowing just one percent of the inner radioactivity to escape from one plant would put as much harmful contamination directly into the environment as 10 bombs. And it would not be spread out all over the globe like bomb fallout; it would all be concentrated in just a few states. Suppose we had to abandon large sections of this land we love?

I can not deny that the government should be preventing this extraordinary possibility. But when we observe that the government allowed harmful conditions to develop in our air and water from other pollutants, then it is clear that citizens had better not count on the government to prevent *nuclear* pollution for them either.

I believe that citizens should get very active, very loudly, very fast.

I am not at all impressed by promises that growing nuclear activities will *never* give us more than a tiny part of the legally permissible radiation dose. Sincerity

XXVI

would require backing those promises with action — like supporting a reduction in the legally permissible radiation dose. I haven't seen that happening. Instead, I read testimony presented by AEC Commissioners to Congress a year ago opposing any reduction at all because, they explained, they did not know how near to the full limit the nuclear power plants might go.

Also, I am not at all convinced by claims that nuclear power plants are safe and that radioactivity will never escape accidentally. If they are as safe as utilities claim, then why did those same utilities insist that they be given special limits on their liability for accidents? When utilities back their claims by supporting repeal of their special liability privileges, I will be more impressed.

It turns out that the government — the Atomic Energy Commission—emphatically does not share the industry's proclaimed confidence that their commercial nuclear-plant designs incorporate adequate provisions for safety. The Atomic Energy Commission (AEC) testified to Congress last year that, "Many safety issues remain to be resolved before a substantial number of these plants (which have received construction licenses) will be able to be licensed for operation." [1]

It was hard for me to believe that the essential system which stands between the public and a radioactive calamity—a system called the Emergency Core Cooling System—has never once been actually tested to see if all parts will work in reality as well as theory. That's something like allowing commercial airlines to use planes which have never been test-flown. Apparently the AEC thinks those emergency-system tests are important, because the Commission is spending many millions of dollars to prepare the tests—for 1975[2].

"Though we can generally tell when we have a very *un*safe (nuclear) reactor, it's always hard to know how safe you are with one you believe to be safe." That statement was part of sworn testimony last year from N. J. Palladino, a member of the AEC's Advisory Committee on Reactor Safeguards, and Dean of the College of Engineering at Pennsylvania State University[3].

Who could fail to be alarmed after carefully reading the following testimony presented last year before Congress?

"There continues to be increasing recognition throughout the nuclear power industry of the urgent needs for development and adoption of engineering standards and other disciplined quality assurance practices . . . In spite of progress, the actions and accom-

---

1. From "Nuclear Safety Program," written testimony submitted by Milton Shaw, Director, AEC Division of Reactor Development and Technology; published in "AEC Authorizing Legislation Fiscal Year 1971," hearings before the Joint Committee on Atomic Energy, March 11, 1970, Part 3, page 1374.
2. From Milton Shaw's testimony, pages 1339, 1363-67. See note #1.
3. From sworn testimony (during question-and-answer period) by N. J. Palladino, before the Select Committee on Nuclear Electricity Generation, Pennsylvania State Senate (Sen. Edwin G. Holl), Harrisburg, Penna., August 21, 1970.

plishments fall far short of what is needed," reported the AEC's Division of Reactor Development and Technology[4].

What does that mean? Admiral Hyman Rickover, the AEC's Director for Naval Propulsion (nuclear submarines), clearly stated that the meaning might be disaster. "There is need for utmost care in design, manufacture, installation and operation of complex systems and equipment inherent in this technology. No carelessness can be tolerated anywhere in the entire chain or the results may prove disastrous. Unfortunately, there are many who are not aware of the necessity of this approach. The difficulties you refer to in connection with fabrication of civilian nuclear central power plants, are, I believe, due largely to failure to specify and enforce the required high standards for systems and equipment."[5]

Sloppy work has been reported from reactor construction sites, as well as from earlier stages of reactor-building. The AEC's inspection staff is far too small to keep control, so the builders are allowed to police themselves. They are almost all running far behind schedule, so the temptation to cut corners is certainly there.

If you imagine that the AEC feels responsible for

4. From "Nuclear Power Industry," written testimony submitted by Milton Shaw, pages 1191-2. See note #1.
5. From testimony by Vice Admiral Hyman Rickover, Director, AEC Division of Naval Propulsion; published in "Naval Propulsion Program 1970," hearings before the Joint Committee on Atomic Energy, March 19-20, 1970, pages 96-101.

your safety, you will be surprised to learn that the Commission has specifically abdicated "basic responsibility for safety" to industry. Said AEC Chairman Glenn Seaborg, "Problems in the design, fabrication, and building of nuclear plants can be minimized only by rigorous quality assurance programs, initiated and enforced by top utility management."[6] AEC Commissioner James Ramey has added, "It must never be forgotten that responsibility for safety of the plant rests with the owner or operator. The regulatory groups, no matter how thoroughly they carry out their function, can not provide complete assurance that public health and safety will be adequately protected in a power reactor project."[7]

When 525 members of the National Society of Professional Engineers were polled a year ago, almost 60 percent answered "Yes" when asked whether there is a valid reason for the public to be worried about nuclear plants.[8]

We must stop this gigantic gamble with public safety.

I endorse bills and petitions which would impose a

---

6. Glenn Seaborg, speaking to the 37th Annual Convention of the Edison Electric Institute; quoted in the 1971 Authorization hearings, Part 3, page 1192. See note #1.
7. James T. Ramey in "AEC Authorizing Legislation," hearings before the Joint Committee on Atomic Energy, 1968, Part 3, page 1186.
8. Reported in "INFO," April 1970; "INFO" is a newsletter of the Atomic Industrial Forum, Inc., 850 Third Ave., New York City 10022.

moratorium stopping construction of nuclear power plants. Before any more are built and licensed, we are all entitled to safety-first policies and to straight answers to many questions — including those regarding the presently "permissible dose" of radiation (please see Appendix II). There are already moratorium efforts underway in California, Minnesota, Oregon, and New York City; by the time you read this, there may be others.

My bill in the Senate would create an Energy-Environment Commission to see that we get *safe* sources of electricity; one section of the bill would stop the stampede to nuclear power by repealing the special insurance privileges (which are explained in this book). Other sections would require that we make the dirty coal and oil plants *clean,* which is possible, and would see that we develop ways to get our energy without poisoning the planet. There seem to be lots of possible ways to accomplish that—such as fusion, solar, wind, and geothermal energies, as well as magnetohydrodynamics (MHD) and fuel cells. The real question to decide during a nuclear moratorium is: Do we take our chances with some of the gentle possibilities, or do we rush into a commitment to the *one* technology which may end up contaminating this planet permanently?

I'm optimistic. There is such a fabulous amount of energy renewing itself naturally on earth that, if man learned to tap just a tiny part of it, he could probably make all the electricity he needs without disturbing nature's harmony. Unless we start putting effort into solving energy needs, it's insincere to say that it is electricity *vs.* the environment, or any of the other false choices offered us. I hope that you will consider some of these ideas and proposals, which are included later (please see Appendix III).

You will probably agree with some of the ideas in this book, and disagree with others, and have good ideas of your own. Please let us know by writing; your opinion counts only if you express it.

Don't expect overnight miracles. Very few of my colleagues in Congress are ready yet to take action on this subject. As you know personally, it takes time to move from feeling

> #1. NOT INTERESTED: "Nuclear plants are safe and clean."

to #2. CONCERNED: "Is there really a hazard?"

to #3. DETERMINED: "I will study and find out."

to #4. CONVINCED: "Yes, there is really a hazard."

to #5. READY AND ABLE: "It's time to convince others."

Most members of Congress are somewhere between stage #1 and #2. Many will skip stages #3 and #5 because it is impossible for one human being to become expert in all the subjects of public importance. Don't expect otherwise.

Expect that a member of Congress will move

from stage #2 to stage #4 whenever a significant number of home-state voters, groups, experts, and newspapers assert loudly enough that there is a hazard, and they expect him to do something about it.

Therefore, the effective thing for you to do is to move yourself into stage #5, and then teach others.

You can challenge professional groups, like your state medical association, your state cancer, heart, and birth defects associations, university and high school biology professors, and your state and national representatives, to take public positions on the nuclear issue. If they plead too much ignorance, insist that they have a responsibility to learn, and help them to do so.

In addition, you can start asking the important but presently unanswered questions about nuclear hazards. Ask the people who ought to have those answers to come meet groups in your area. Invite the Atomic Energy Commissioners,[9] members of the Joint Committee on Atomic Energy,[10] the Atomic Industrial Forum,[11] National Committee on Radiation Protec-

---

9. The AEC Commissioners are:
   Glenn Seaborg (chemist).
   James T. Ramey (lawyer).
   Wilfred Johnson (mechanical engineer).
   Clarence E. Larson (bio-chemist).
   Address: U.S. Atomic Energy Commission, Washington, D.C. 20545.
10. Joint Committee on Atomic Energy, U.S. Capitol Bldg., Washington, D.C. 20510. The Committee can provide a list of its new membership for the 92nd Congress.
11. The Atomic Industrial Forum's membership comprises about 600 firms and organizations and government agencies engaged in development and utilization of nuclear energy for peaceful purposes. Address: 850 Third Ave., New York City 10022.

tion[12], executives and board members of the utilities, members of your state public utility commission, state engineering board, and university engineering faculty.

You may wonder why I suggest invitations to people who will favor nuclear power. I am not suggesting invitations for them to give long, rosy speeches; I am suggesting invitations for them to answer the hard questions.

These are some of the people we would invite to a Senate hearing. After all, these are the people responsible for bringing us the threat of nuclear accidents and pollution. These are the men who are obliged to explain themselves and to tell us what they know and what they do not know.

The problem is that you must know your subject extremely well before you extend that kind of invitation; otherwise, you will be "snowed" and unable to recognize a false or inadequate answer. It takes lots of knowledge just to ask the right questions.

You can tune in to some of the valuable experience of others who joined the fight earlier.

---

12. The National Committee on Radiation Protection (NCRP) is a non-profit corporation chartered by Congress in 1964. Its work, which includes recommending a permissible radiation dose for approval by the Federal Radiation Council, is supported financially by 33 organizations such as the U.S. Atomic Energy Commission, the U.S. Public Health Service (HEW), the American Nuclear Society, the American College of Radiology, the National Electrical Manufacturers Association, and the Office of Civil Defense. Address: 4201 Connecticut Avenue, N.W., Washington, D.C. 20008.
The NCRP's most recent report is #39, "Basic Radiation Protection Criteria," issued January 15, 1971; price $2.00.

No one loves a complainer.

Citizens who object to something—like radioactive power plants—will find sympathy and quicker success if they also propose a better solution. Safe alternatives to nuclear electricity do exist, and this book will introduce you to some. It is almost hopeless to oppose a nuclear power plant unless you think through and find endorsement for another solution in your area. Students in economics and engineering may be willing to help.

You will each have to make a personal choice: will I work on it, or just hope that others will take care of the problem for me? For make no mistake, there is a *problem*.

Suppose that no one else does do the job for you successfully? The future in a badly contaminated world would be very grim.

Even now, every time a woman enters the maternity ward to have a baby, she faces one chance in twenty-five that she will give birth to a child who is seriously defective mentally or physically. Nuclear pollution would make the odds worse; do American women want that? It is a fact that pregnant women (the embryos they carry) are far more sensitive than anyone else to radiation damage.

Cancer presently kills more children than any other cause except accidents. Nuclear pollution would cause even more children to suffer and die from cancer; do citizens approve of that? Children are considerably more sensitive to radiation harm than adults, but adults also can get cancer from nuclear pollution.

If there were serious nuclear pollution, most healthy people might have to spend much of their lives caring for sick people.

Lasting nuclear pollution — practically permanent radioactive contamination of this planet—is a possibility and even a probability unless some present policies are changed soon. Therefore, I believe our descendants will not forgive faint-hearted efforts from any of us.

> —*Mike Gravel, U.S. Senator from Alaska*
> *Washington, D.C.*
> *March 1, 1971.*

Introduction:

The Nuclear Juggernaut

Look back a few short years ago upon the construction and attempted operation of the so-called "Fermi" nuclear power plant, some 30 miles from the heart of Detroit. The events leading up to that fiasco are most instructive. We quote Bryerton* on this:

"The Advisory Committee on Reactor Safeguards, a panel established by Congress to advise the AEC, reported to then-chairman of the AEC, Lewis Strauss, on June 6, 1946, '. . . the committee believes there is insufficient information available at this time to give assurance that the PRDC (Power Reactor Development Corporation) reactor can be operated at this site without public hazard.' Strauss suppressed the

---

*"Nuclear Dilemma." Gene Bryerton, Friends of the Earth/Ballantine, New York, 1970.

safeguards committee report, and less than two months later the AEC gave its approval for issuing a construction permit."

After a series of court battles, the U.S. Supreme Court finally permitted the construction of this nuclear power plant. In a dissenting opinion, Justices Douglas and Black stated:

"The construction given the (Atomic Energy) Act by the (Atomic Energy) Commission is, with all deference, a lighthearted approach to the most awesome, the most deadly, the most dangerous process that man has ever conceived."

But such words of wisdom and caution are routinely dismissed by the "pioneering" champions of the nuclear juggernaut.

The sad history of the Fermi plant is now well known to everyone. After an outlay of 120 million dollars this ill-starred atomic boondoggle failed on October 5, 1966, in its early stage of operation. A period followed where no one involved knew whether the entire city of Detroit would have to be evacuated. We were lucky that such a secondary disaster did not follow the failure of the Fermi reactor. We emphasize the word 'lucky', and no one has expressed this better than Professor Edward Teller, who, by the way, is a fervent *proponent* of nuclear energy.

". . . So far, we have been extremely lucky. . . . But with the spread of industrialization, with the greater number of simians monkeying around with things they do not completely understand, sooner or later a fool will prove greater than the proof even in a foolproof system."**

Yes, a nuclear juggernaut, responsive to no societal needs, has been moving across the land, threatening a nightmare for life on earth—forever. Having come to appreciate the meaning and peril of the mad stampede to nuclear power generation, we feel it is urgent that all people understand now what they are facing, so that constructive, positive action can be taken before it is too late. Those sponsoring this rash and dangerous industry are desperately anxious that the true facts be withheld from the public.

It is our purpose to explain how nuclear electricity generation may imperil your health, your life, or your property, without any possibility for redress. You must learn that this stampede is being foisted upon you undemocratically as it proceeds rashly and unwisely to put all our major metropolitan centers at serious risk. Most importantly, you must learn that citizens do indeed have the power to stop this juggernaut and thereby protect their lives, their children's lives, and their property. You cannot count on government, industry, or anyone else; *you* must learn to act effectively. The public has been and *is being* deceived by a clever, well-financed propaganda campaign of delusions concerning, "clean, cheap, safe nuclear power."

**As seen in the *Eugene Register Guard* (Oregon), October 7, 1969.

3

Nuclear power has *not* been proved to be clean!
Nuclear power has *not* been proved to be cheap!
Nuclear power has *not* been proved to be safe!

We believe your consideration of all the evidence will lead you to agree that the nuclear fission approach to electricity generation, represents the worst possible choice available to us. The evidence is not at all complicated. It encompasses three major areas:

(1) The *poison* (radioactivity) that is an inevitable by-product.

(2) The unknown safety of the *power* generation itself.

(3) The attractive and essential *alternatives* available now and in the future.

THE POISON

Radioactivity represents one of the worst, maybe the worst of all poisons. And it is manufactured in astronomical quantities as an inevitable by-product of nuclear electricity generation. One year of operation of a single, large nuclear power plant, generates as much of long-persisting radioactive poisons as *one thousand* Hiroshima-type atomic bombs. There is no way the electric power can be generated in nuclear plants without generating the radioactive poisons. Once any of these radioactive poisons are released to the environment, and this we believe is likely to occur, the pollution of our environment is irreversible. They will be with us for centuries. It is important that people learn

4

how they are likely to be exposed to such poisons and *how* death-dealing injury is thereby produced in the individual and in all future generations.

THE POWER

Careful examination of the generation of electric power by nuclear generators is essential, for such examination shows that, even in its infancy, this industry might have caused a major health calamity. We have been lucky thus far, and we have no assurance whatever that our luck will hold out, as this nuclear generating industry burgeons rashly and unwisely. The leadership of this nuclear juggernaut has been anything but responsible. The top leadership has displayed a total lack of comprehension of radioactive poison and its effects. The top leadership has displayed a total lack of understanding of the basic principles of *sound* public health practice which must be applied to new, hazardous technologies. Worst of all, this same top leadership has demonstrated a lack of responsibility in meeting the moral obligation to provide the public with honest information concerning the real hazards which must be faced. We are not speaking of usual "industrial accidents." Rather, we are concerned over the hazard of major calamities to human health and life, unparalleled in human history.

THE ALTERNATIVES

The nuclear juggernaut is by no means necessary to guarantee us an adequate supply of electric power.

Far from it. There are several attractive, feasible alternatives to meet electric power requirements. The choice of a rational alternative is in the hands of the public. Citizens undoubtedly can and should exercise their power to choose a rational viable future. Constructive action is possible to protect lives, health and property. It is important to learn how such constructive application of citizen power can be utilized to counter the nuclear juggernaut.

THE MASSIVE HOAX

About a year ago, we began to perceive the dimensions of the massive hoax being perpetrated upon the public. It was very difficult for us to believe that what we observed to be occurring could truly be real. Indeed, up to that time we, deeply immersed in atomic energy research, had been lulled into the belief that nuclear electricity was the one atomic energy program which posed very little threat to society. How wrong we were. There is a real potential disaster ahead, we now know. It is, we think, important that you share this knowledge with us.

In 1963 we were asked by the Atomic Energy Commission to undertake long-range studies of the potential dangers for man and other species from a variety of so-called "peaceful uses of the atom." Nuclear electricity generation is one such atomic program. Naturally, we presumed that the Atomic Energy Commission seriously wanted to know the truth concerning the magnitude of possible hazards. In fact, in assigning this study mission to us, Chairman Glenn Seaborg assured us that he wanted favorable or unfavorable findings made available to the public. "All we want is the truth," Chairman Seaborg said in 1963.

We have learned, to our great dismay, that these assurances were illusory. It is now clear to us that the Atomic Energy Commission had not contemplated seriously that the studies might reveal serious flaws and dangers in the "peaceful atom" programs. This kind of "truth" has proved to be quite unwelcome. The research findings we have made can be expressed succinctly in two statements:

(1) Radiation, to be expected from several atomic energy programs burgeoning rapidly, is a far, far more serious hazard to humans than any of the so-called "experts" had previously thought possible.

(2) The hazard to this generation of humans from cancer and leukemia as a result of atomic radiation is TWENTY TIMES as great as had been thought previously. The hazard to all future generations in the form of genetic damage and deaths, had been *underestimated* even more seriously.

In a rational society, where the health and welfare of citizens would be considered as paramount, such research findings would have been warnings welcomed. And we expected our findings would be welcomed. Instead we received a torrent of vitriol and personal condemnation from three major sources:

The U.S. Atomic Energy Commission
The Joint Committee on Atomic Energy

7

The Electric Utility Industry

What, we wondered, could account for the reflex attack upon us by these three groups. What made these *particular* three groups so inordinately sensitive to new knowledge concerning the hazard of radiation? What do these three groups share as a common interest? The answer was not far to seek. These three groups act in concert as a powerful, promotional triumvirate for nuclear electricity generation. It was not immediately obvious to us why, even so, such a triumvirate should be so violently frightened by knowledge concerning radiation hazards. In fact, one might logically believe knowledge important for the nuclear industry's benefit would be an item of extremely high priority.

Slowly it dawned upon us that possibly all was not so wonderful about nuclear electricity generation as the advertisements claimed. Could it be that this triumvirate had been misleading the public, lulling the public into the belief that no significant hazard existed? Was something being swept under the rug? A little careful probing showed that not only was *something* being swept under the rug; the entire structure was rotten in every respect.

This entire nuclear electricity industry had been developing under a set of totally false illusions of safety and economy. Not only was there a total lack of appreciation of the hazards of radiation for man, but there was a total absence of candor concerning the hazard of serious accidents. The economics were being treated with rose-colored glasses. And the triumvirate knew all too well that the stampede to nuclear power, initiated by them, could not possibly tolerate the bright light of exposure to public scrutiny. The more deeply we probed, the more we realized how massive the deception truly was. It became quite clear that concealment of truth from the public was regarded as essential. And we realized very clearly why the violent reaction had greeted our presentation of the research findings concerning the radiation hazard to humans.

Further, we came to understand why citizen groups, across the country, were becoming increasingly concerned and alarmed about the plans for ringing essentially every major metropolitan center with gigantic, totally experimental, untried nuclear power plants. And such public groups, expressing their apprehension, didn't even know all the extremely serious hazards of radiation about which we had learned through our researches. By speaking with such concerned groups and by listening to them, we finally came to understand that they had found no sincerity, no honesty, and no candor being displayed by the Nuclear Triumvirate with respect to provision of the real facts behind the nuclear power story. Pure public relations department rubbish was being passed out to the public, heralded as "information." The false illusion that citizens with concern could be "heard" was really a mockery of a democratic tradition. Interventions in Hearings before AEC Licensing Boards were a joke. The Licensing Boards, chosen by the AEC, and known to be favorable to nuclear power development, went through the motions of a Hearing and inevitably returned a rubber-stamp decision that a go-ahead should be given to every nuclear power plant.

The promoters of nuclear electricity have recently been bemoaning the public "alarm" and have been criticizing those who raise questions as "stirrer-uppers" or as conservationist "kooks." We do not for one moment worry at all about the *public's alarm* over the dangers it faces from an irresponsible nuclear juggernaut. The real source of concern is that the public is not *sufficiently* alarmed and frightened over the rash, irresponsible actions of the men who are stampeding this country into a potential nightmare. These men are much more frightening than the hazard of radiation.

How shall we cope with men who initiate and promote rash technologies that can spell irrevocable disaster? It is truly unfortunate that the men responsible for ill-advised technological ventures, such as nuclear electric power generation, are rarely called upon *by name* to explain their actions. The cloak of anonymity protects them from responsibility for what they do, even if their actions ultimately compromise the very survival of humans.

It is of the utmost importance that every concerned citizen learn to identify the men behind anti-social and anti-human, ill-considered technological ventures. It is of the utmost importance that the men who promote such industries be identified publicly and be asked by the public to answer the serious moral questions they have thus far sidestepped. One cannot ask a nameless bureaucracy anything. But the men within the bureaucracies must be brought forward to answer the questions they seek to avoid.

Congressman Chet Holifield, a member of the Joint Committee on Atomic Energy, sits at the very top of the atomic energy pyramid. He, a super-hawk promoter of nuclear energy, continues to make public statements verbally, and on the printed record, which display an appalling ignorance of even the most elemental aspects of radiation hazards for humans. Amounts of radiation exposure that concern the world biological community as being characterized by potentially disastrous long-term consequences, are described by Mr. Holifield as "abundantly safe." It is not that Mr. Holifield hasn't been provided the necessary biological and medical information. He simply displays no interest in listening to such information.

The U.S. Atomic Energy Commissioners,* currently Seaborg, Johnson, Ramey and Larson, provide a steady stream of platitudes concerning nuclear power, platitudes that do not even remotely address the real issues of concern. Instead of facing their moral and social obligation to explain truthfully their total ignorance of the hazards of major accidents in nuclear power plants and of sabotage at many steps throughout the nuclear power industry, they simply issue soothing unsupported

---

*It was with profound sorrow that we learned of the untimely accidental death of AEC Commissioner Theos Thompson after this book was written. While many of the comments made by Commissioner Thompson concerning radiation hazards are criticized in the text of this book, such criticism in no way was ever directed toward the person of Commissioner Thompson.

and unsupportable statements concerning safety of such plants and operations. None of the Commissioners has even begun to address himself to the serious questions raised repeatedly by scientists and by the public-at-large. Their major recent answer to serious public concern has been to double the size of their public relations efforts—all generously supported at taxpayer expense. The public must invite these men to begin expressing, in a responsible manner, the real questions surrounding their nuclear power promotions.

Lastly, we must turn our attention to the directors and presidents of the electric utility industry. The names of these men are rarely known to the public. They shun public questioning concerning the serious moral issues they have failed to face squarely. Without any valid evidence their publicity departments say we *must* have nuclear power to meet our electricity needs. At the very same time they authorize the spending of millions of dollars to advertise for increased *consumption* of electric power. No better scheme for creation of self-fulfilling prophecies could be imagined. And this, too, is of course ultimately added to the consumers' electric bills.

Further, the directors and presidents of electric utility industries staunchly refuse to learn anything concerning the true hazard of their by-product radioactive poisons. It is regrettable that initially they were duped into believing no hazard existed, through the falsely optimistic statements emanating from the Atomic Energy Commission. This, however, cannot be accepted as an adequate basis for their steadfast refusal to learn.

SOME VALUABLE CLUES

If one is desirous of knowing about environmental hazards, there exists *one* unusually reliable source of unbiased information. This is the insurance industry. This highly successful industry has consistently operated profitably for one reason—they understand how to assess hazards. For the insurance industry to fail in their assessment of hazards means the loss of money. The insurance companies do not lose money.

One has only to observe the actions of the insurance companies to realize that this industry has essentially no confidence in the safety of nuclear electricity generation. The insurance people have watched carefully the burgeoning of nuclear electric power and they wasted no time taking serious steps to protect themselves. Little known to an unsuspecting public, the insurance industry inserted exclusion clauses into Homeowners' policies. If homes and property suffer damage from radioactivity and nuclear plant accidents, most policies do *not* cover such damages. Many people have thought such exclusion clauses apply to nuclear war. Far from it—they are specifically planned to protect the insurance industry against the nuclear power industry. In this manner the profit-minded insurance companies have looked at nuclear electric power and have expressed themselves, "No Confidence."

And what about the bulk of the liability for *major*

damage from nuclear power plant accidents? Who carries this? No one—no one at all. The electric utility industry, of course, would never have even considered venturing into nuclear power if they could be held liable for the disastrous accidents considered possible. The insurance industry refused to insure such risks at any price. The Congress, with flagrant disregard for the public's rights, absolved everyone of virtually all significant liability by passing a law, known as the Price-Anderson Act. A major accident, it has been estimated, could lead to 7 billion dollars worth of damage. The Price-Anderson Act decrees that the maximum liability shall be 560 million dollars. The public stands to recover a *maximum* of seven cents on each dollar in such an event. And the public pays 80% even of this inadequate protection.

If the electric utility corporations had to stand financially behind their sweet assurances of "safe" nuclear power, the nuclear electricity industry would cease immediately.

WHAT YOU MUST DO

The nuclear juggernaut plans fully to roll on without assuming any of the moral and social responsibility it should take. So long as the public is lulled to sleep and doesn't know the right questions and the necessary actions, this madness can and will go unchecked. It is a major purpose of our book to provide the information concerning the supremely important questions the nuclear electricity spokesmen have failed to answer.

An *informed* public can act effectively to protect itself. Such *action* is essential.

Too many people feel the problems and the questions are highly technical. And because of this they are easily fooled into believing they must rely on the "experts." But a new day of awareness is dawning and a large segment of the public is fully aware that it has been the "experts" who have time and again misled us and have brought us to the sorry environmental plight we face today. Be your *own* expert. You *can* understand every aspect of the problems created by the nuclear juggernaut. We believe the ensuing chapters will provide you with the knowledge required for taking intelligent, effective action.

You may find a particular point *seems* technical, momentarily. We can assure you that you can pass such points by and go right on without any loss of understanding the real meat of the problem. At a later time you will find it easy to return to and to understand those few issues you went over rapidly. What you must realize is that if you do not have the information at your disposal, or do not know where that information is to be found, you will be a sitting duck for falsehoods and irrelevancies that will be thrown at you. Honest information lies at the root of strength for public action to protect the environment, your property, your health and your life! But you don't need to understand each little detail, provided you can get it when you do need it.

WE ARE OPTIMISTIC

We have no desire whatever to be prophets of doom concerning the serious threat posed to you by the nuclear electricity industry. Not at all! We know that the public can and will take care of this mad stampede very effectively. An aroused and informed public is the key to self-defense against environmental rapists.

Therefore, after describing the nature of the poison generated by the nuclear electricity industry and the enormous price in health and property this industry can cost the public, we turn to the really important issues to be treated here. First you will learn that there are highly attractive alternatives available which make nuclear electric power generation unnecessary. You will realize there is no basis for concern that stopping the nuclear juggernaut means a shortage of electric power. We *can* have the electric power we need. We can have a high quality of life. And we can and must stop the destruction of our environment by the introduction of poisons that will last essentially forever.

People have underestimated their power and ability with respect to turning back polluters. Truth and information spread rapidly. There is nothing that the Atomic Energy Commission, the Joint Committee on Atomic Energy, and the electric utility industry fear more than an informed public. An informed public can act, and act with great effectiveness.

And if stopping the nuclear juggernaut is accomplished through intelligent public action, the gain will spread. A successful attack upon this particular ill-conceived and rash stupidity of man will give the public confidence and the tools to solve other problems threatening our environment and our planet.

We are quite optimistic that the public can teach the parochial, shortsighted nuclear triumvirate that the planet Earth has other functions than to serve as a sewer for the triumvirate's wastes.

CHAPTER 1

Nuclear Reactors to Generate Electricity

The fundamental difference between a nuclear electrical power plant and a conventional power plant is the fuel that is employed. The conventional plants burn coal, oil, or gas to create the heat while the present nuclear plants burn uranium. Burning *1 ounce of uranium* has roughly the same potential as burning *100 tons of coal*. A ton of reactor fuel may substitute for many loaded freight trains of coal.

The purist may resent the choice of words, "burning uranium" because the mechanism is quite different from ordinary combustion. The burning of fossil fuels such as coal results from the carbon combining with oxygen to form carbon dioxide with the release of heat. The burning of uranium results from the uranium combining with an atomic particle called a neutron and subsequently splitting into lighter elements such as strontium and cesium with the release of a large amount of heat. This process is called "nuclear fission" rather than burning.

Many lighter elements that are formed following the fission of uranium are radioactive. They are the same radioactive elements that caused so much concern over the fallout from the atmospheric testing of nuclear weapons. It is these deadly radioactive substances that represent the unique and serious hazard from nuclear power plants. We shall have much to say about these materials in the subsequent chapters.

One may wonder why fission power is being considered at all if it represents a serious hazard. There are two reasons. First, the world's supply of fossil fuels is finite and eventually man will have to find another source of energy. Second, and this seems to have been the compelling reason for the present rash proliferation of nuclear power reactors, the cost of nuclear electrical power was projected to be cheaper than that from conventional plants. The first reason is justifiable, although as we shall indicate later, it is not of immediate importance. The second reason and the basis of the present rash proliferation was wrong. *Nuclear power is more expensive.*

Although there are a variety of nuclear reactor designs that are possible and in existence, the major choice in this country has been made in favor of the water-moderated reactor and the liquid metal fast-breeder reactor.

Water-Moderated Reactors

When a uranium atom absorbs a neutron and undergoes fission, in addition to producing the two lighter elements, it also emits two or three neutrons. These neutrons can, in turn, react with other uranium atoms, which will undergo fission producing more neutrons. Since more neutrons are produced than consumed, it is possible, under the right circumstances for this reaction to proceed at an ever increasing rate and produce a tremendous amount of heat in a very short period of time. In other words, an uncontrolled nuclear reactor could literally blow itself up. Controlling a nuclear reactor therefore, means controlling the mutiplication of neutrons in the reactor core.

The reactor core is a cylindrical steel containment vessel into which the fuel elements are inserted. The fuel elements are an assemblage of long slender rods that contain the uranium in the form of an oxide. Depending upon the size of the reactor, it will contain a large number of these elements and hence thousands of fuel rods. The geometric arrangement of these rods and elements is important because the chain reaction depends upon having a particular concentration of fissionable material in a particular volume. For example, a certain amount of fissionable material in a particularly shaped vessel can be perfectly safe. If, however, it is put into a vessel of a different shape, the mass can become critical and the chain reaction can take place with explosive intensity.

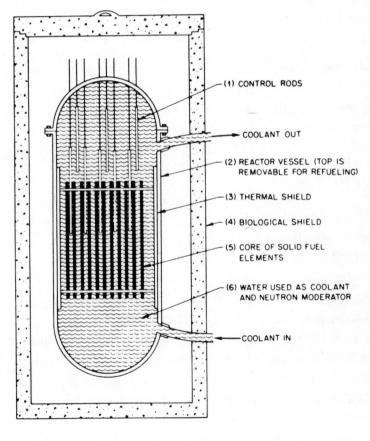

(1) CONTROL RODS

COOLANT OUT

(2) REACTOR VESSEL (TOP IS REMOVABLE FOR REFUELING)

(3) THERMAL SHIELD

(4) BIOLOGICAL SHIELD

(5) CORE OF SOLID FUEL ELEMENTS

(6) WATER USED AS COOLANT AND NEUTRON MODERATOR

COOLANT IN

Vertical cross-section of a pressurized-water nuclear reactor.

Once a reactor core is assembled, it has the critical mass of uranium in an appropriate volume. That is precisely why the reactor works. But in addition to the critical mass of uranium, the reactor has control rods. These rods when inserted into the reactor core are able to absorb neutrons. When they are all inserted into the core, they absorb so many neutrons that not enough are available to sustain the chain reaction. As the rods are gradually withdrawn the power level of the reactor increases.

The above explanation of controlling a nuclear reactor sounds simple. In principle it is simple. But between that simple explanation and the design of a truly safe reactor lies a great deal of engineering sophistication. We shall go into this further in Chapter VI.

Besides safety, there is a collateral aspect of nuclear power plants—i.e. reliability. Because of the serious hazard associated with the radioactivity accumulated in the core of a reactor, the safe operational limits of the reactor relate to its reliability. A nuclear power station may be required to shut down simply because it is releasing, or may potentially release, too much radioactivity to the environment. One of the questions that we explore in this book is—if adequate (more restrictive) regulations were imposed on the present nuclear power reactors would they be allowed to operate at all?

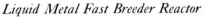

Liquid Metal Fast Breeder Reactor

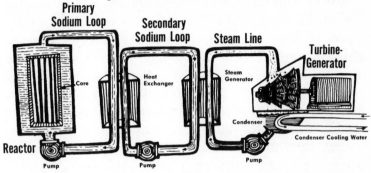

Pressurized Water Reactor

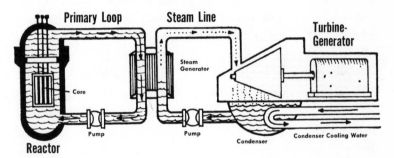

Breeder Reactors

The fissionable material in the present water-moderated reactors is uranium-235 (U-235).

Uranium-235 represents only about 1 percent of the natural uranium (0.71 percent). The rest is composed of the heavier isotope U-238. Uranium-238 can not be made to undergo fission except by high energy neutrons which are not created when U-235 undergoes fission. However, U-238 can be converted into a fissionable material, plutonium-239 (Pu-239) when it absorbs a neutron.

The present-day nuclear reactors discussed above are moderated by water. By moderated it is meant that the neutrons which the U-235 releases upon undergoing fission are slowed down (reduced in energy) by the water. The present nuclear reactors also contain U-238 and when the U-238 captures a neutron it is converted to Pu-239. In the present reactors, because the neutrons are moderated or slowed by the water in the reactor, the U-235 emits fewer neutrons than it would if it underwent fission as a result of absorbing faster neutrons. As a consequence, less Pu-239 is made in the present reactors than U-235 that is burned. There is a net consumption of fissionable material; considerably more fissionable material is consumed than is produced.

In the fast-breeder reactors, however, the water moderator is removed and the neutrons which are captured in this case by the U-235 are faster, or higher energy, neutrons. The net result of this is that the U-235, when it undergoes fast neutron fission, produces more neutrons than it would when it undergoes fission as a result of the absorption of a slow neutron. This production of additional neutrons results in a net increase in the production of plutonium over the amount of U-235 that is consumed. That is why these reactors are called breeders. The operation of the fast-breeder reactor, therefore, is anticipated to produce large quantities of fuel to operate reactors fueled by Plutonium-239 (Pu-239) rather than the present uranium-fueled reactors.

But the requirement for higher energy neutrons in order to produce a breeder reactor means that the reactor will necessarily have to operate at a higher temperature. The neutron moderator, the water, is therefore not used in these reactors and the reactors are cooled with the liquid sodium. Sodium is a highly reactive metal and liquid sodium will explode upon contact with water or air. Moreover, the fast breeders, in order to increase their breeding capacity, have to be made more compact than the present generation of reactors. Therefore, they will contain much more fissionable material in a smaller volume.

The breeder reactor, therefore, concentrates the fissionable material into a smaller volume and operates at a significantly higher temperature. These two specifications for a fast-breeder reactor represent the most serious engineering complications for these systems. The major concern with these reactors is an accident that might result in the concentration of fissionable materials into small volumes wherein the chain reaction can proceed in an unmoderated fashion. Such an event could result in an extreme increase in temperature and a possible explosion. These explosions are not as tremendous as those which result from atomic bombs which are designed for this particular purpose, but nevertheless it is these potential explosions which represent the grave concern of nuclear-reactor designers. Consequently, the fast-breeder reactor places the most stringent requirements upon the control of the process to prevent over-heating and melting of the fuel materials. Very small melts can result in the accumulation of critical masses.

Again it may be asked: why proceed with the fast-breeder reactor if it is potentially dangerous? One reason is that the present-day water-moderated reactor will rapidly consume all of the U-235, and nuclear reactors would disappear. The breeder is an answer to this. It will produce more fissionable material in the form of Pu-239 than it uses. The other reason for developing the breeder is that it was envisioned as a source of very cheap power. However, as we shall show later, this vision is becoming quite blurred.

There is considerable reason to suggest that the Atomic Energy Commission made a serious mistake some 15 years ago when it began to press for the rapid development of a nuclear power industry. The first chairman of the AEC, David Lilienthal, has recently stated: *

> Once a bright hope, shared by all mankind, including myself, the rash proliferation of atomic power plants has become one of the ugliest clouds overhanging America.

---

*David Lilienthal, New York Times, July 20, 1969.

We share Mr. Lilienthal's apprehension and much of this book explains the basis for our concern.

The last three chapters discuss the methods available and the means required to rectify this mistake which the AEC seems bent on perpetuating.

Suggested Reading

A more complete description of nuclear power plants, together with pictures and drawings, can be obtained in three booklets published by the Atomic Energy Commission in its "Understanding the Atom" series:

Nuclear Reactors

Nuclear Power Plants

Atomic Power Safety

These three booklets may be obtained free by writing to USAEC, P.O. Box 62, Oak Ridge, Tennessee 37830.

CHAPTER 2

How Radiation From Atomic Energy Programs Gets to You— What it Does to You

To understand why there is grave concern about nuclear electricity generation, it is necessary to know just how nuclear electric power exposes human beings and other living things to the danger of being irradiated. And, it is essential to understand a few extremely simple points about radioactivity, if we are to bypass the confusion on safety points generated by well-planned propaganda campaigns.

The nuclear fission of uranium or plutonium releases enormous energy. This is the energy, ultimately converted to heat, which produces steam to drive turbines and produce electricity. If this were the only energy released when uranium or plutonium atoms split, nuclear electricity might well have been a real boon to mankind, providing electric power for decades or centuries. Unfortunately, the fissioning itself is only the beginning. There is another source of energy involved—after the fissioning is completed—a source which creates the extreme radiation hazard that accompanies nuclear electricity generation.

The uranium itself decays or disintegrates very slowly. This is why uranium is still present on earth so long after the earth was formed. Uranium is radioactive, but only feebly so. The kind of uranium (uranium-235) which maintains the chain reaction in a nuclear reactor decays by emitting so-called alpha particles very slowly, so slowly that it takes 710 million years for one-half of the uranium-235 atoms to disintegrate. For any radioactive substance, the time required for one-half of it to disintegrate is called the half-life. Thus, uranium-235 has an extremely long half-life and, consequently, the radioactive hazard of uranium-235 is very small.

When a uranium-235 nucleus splits, two (usually) "fission fragments" of the original nucleus are produced. What are these "fission fragments"? They represent what is left after a neutron has disintegrated the uranium-235 nucleus and after some additional neutrons have escaped during the fission process itself. It has been discovered that these fragments are nothing more or less than variant forms of elements commonly occurring in nature.

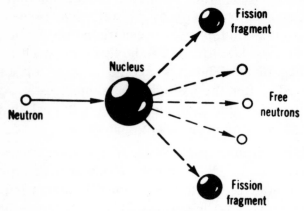

A fission reaction.

The vast majority of the variant forms of the elements found in the earth's crust are referred to as *stable*. This means that they do not decay radioactively or create radiation as does uranium.

The variant forms of any particular element are called *nuclides* of that element. So, we can say that in nature we find several *stable* nuclides for one particular element.

The fissioning of a uranium-235 nucleus can and does produce stable nuclides. But it also produces radioactive nuclides of many chemical elements. Any particular uranium nucleus has a certain probability of splitting, or fissioning, to produce any one, of many hundreds, of stable nuclides or radionuclides, of many different chemical elements. Thus, while one cannot predict which stable nuclide or radionuclide will be produced by the splitting of a *particular* uranium

30

nucleus, we do know with great accuracy, now, how many times out of a million uranium fissions a radionuclide—for example, tin-123 or iodine-131—will result.

Remember that the stable nuclides produced, when the uranium atom splits, are the same as stable nuclides already present in nature. Thus, biologically they create no problem whatever. But, the radionuclides produced at the same time, represent most unwelcome by-products of the operation of nuclear reactors for electric power generation.

So long as the radionuclides remain in the nuclear reactor, they create no problem. True, some of the decaying radionuclides within the reactor do produce gamma rays which can penetrate through steel or concrete, but sufficient shielding and distance can reduce any risk from this source to a negligible level. But, should such radionuclides get outside the reactor —they can and do so—a host of problems can ensue. The problems arise because humans and other living things can be irradiated by them.

A human can be irradiated by such radionuclides from outside his body (external radiation), or from inside his body (internal radiation). There is no difference in biological effect, whether the radiation is external or internal, *provided* the same amount of radiation is absorbed in a particular part of the body.

The common forms of radiation resulting from uranium fission all have the same effect upon living tissue which absorbs them.

It also makes no difference, biologically, whether the tissue receives a specific amount of radiation, in a given time period, from a radioactive particle of long half-life (like thousands of years) or from one of short half-life (like a few days).

Knowing these basic facts, we can see that the only real issue that concerns us, about any radioactive material, is how much radiation is absorbed by a single portion of tissue plus the length of time in which this absorption occurs. In order to systematize all radiation measures, radiant energy absorption is measured by the RAD unit. A substance is said to receive 1 RAD of radiation when it absorbs a given amount of energy per gram of that substance.

One rad delivered to a piece of muscle will have essentially the same effect whether it comes from a dental x-ray machine, from radioactive atmosphere irradiating the muscle from the outside, from gamma rays from radioactive substances in the earth, or from radioactive substances taken into the body with food or water. And it matters not at all, whether the source of radiation is a nuclear power reactor or any other source. *One rad* is *one rad* for nearly all types of radiation, from any source for all practical purposes.

If only the rads that are absorbed matter, why bother about external versus internal radiation, or about the differences from one radioactive element to another?

(a) External versus internal radiation

The various radioactive elements produced by

uranium fission emit beta rays of varying speeds. The speed determines how far the beta ray can penetrate into the body. Thus, a radionuclide may emit only very low-speed beta rays that don't even penetrate beyond the deep layers of the skin. The internal organs are safe when such a radionuclide is placed outside the body. However, the damage to skin from such an external source can be severe.

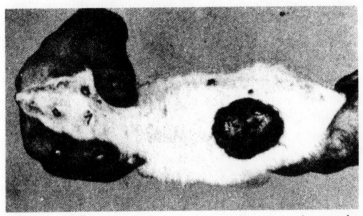

This rat is one of a group of experimental rats that were each exposed to a single dose of beta rays. Six months after irradiation, the group of rats began to develop skin tumors. Tumors were still appearing at 14 months after the single dose of beta rays. Chance of tumor rises with radiation dose.

On the other hand, if this same radionuclide were ingested with radioactively contaminated food or water and distributed uniformly throughout the body tissues, *all* body organs can be irradiated by it.

The case is the same for x-rays or gamma rays. Low-energy x-rays or gamma rays from outside the body may not affect a deeply-situated internal organ.

On the other hand, if the radionuclide emitting those same x-rays or gamma rays is taken into the body, it can affect any organ its x-rays or gamma rays reach.

(b) Concentration in organs

We must recall that the radioactive elements produced in a nuclear reactor behave almost precisely as do their non-radioactive counterparts. For example, radioactive iodine-131 behaves chemically and biologically just as does stable, or non-radioactive, iodine.

But chemical elements do differ from each other in how they distribute themselves once they are inside the body. Iodine is interesting because the thyroid gland has a special affinity for iodine. As a result, the thyroid accumulates far, far more iodine from an ingested dose than any other body organ does. The thyroid uses iodine to manufacture its major active hormone, thyroxin.

The radioactive forms of iodine (iodine-131 is one) behave just as non-radioactive iodine would when taken in with food, accumulating preferentially in the thyroid gland. As a result, that tissue receives a far higher radiation dosage in rads from the decaying radioactive iodine than other tissues of the body do. Naturally, radioactive iodine has its major biological effects on the thyroid gland, compared with its effect on other cells in the whole body.

There is nothing special about the radioactivity of iodine-131 that makes the thyroid gland vulnerable. Precisely the same injurious effect to the thyroid gland could come from x-rays out of an x-ray machine, or from a source of external radioactive cobalt. Only the rads that accumulate in the thyroid gland cells during a particular time matter in assessing possible damage. If half the radiation comes from the outside (as from an x-ray machine) and half from ingested iodine-131 the injury to thyroid cells will be the sum of the rads delivered by both radiation sources.

Certain chemical elements taken into the body along with food or water do not concentrate in specific tissues or organs. Instead, they distribute themselves throughout the body. Cesium, produced abundantly in its radioactive form, cesium-137, by uranium fission, does precisely this. It is said to produce whole-body radiation. Again, the radiations from cesium-137, radioactive iodine-131, or any other radionuclide, or x-rays from a machine are comparable. The effect upon cells anywhere in the body depends solely upon the number of rads they absorb in a particular time period.

(c) The role of chemical similarity between elements

Certain chemical elements are grouped together because they are particularly similar to each other in chemical (and, hence, biological) properties. Lithium, sodium, potassium, rubidium, and cesium represent one such chemical group. Though these elements are by no means *identical* in chemical or biological behavior, they do show numerous marked similarities in properties, including the way living things use them in their bodies.

Potassium is a prominent, vital constituent of the

interior of every living human cell. Fish living in fresh water, where the concentration of potassium is very low, may be forced to concentrate the potassium 1000-fold, in order to maintain the concentration necessary to sustain life. Because cesium is chemically quite similar to potassium, the same mechanism also concentrates cesium from such a fresh water source approximately 1000 times. If the cesium in the fresh water happens to be radioactive cesium-137, from a nuclear reactor or other source, then the fish will contain 1000 times more cesium-137 than the fresh water itself, on a weight-for-weight basis.

So here is an illustration of how the remarkable similarity in behavior of certain chemical elements can lead to massive concentration of radioactive substances in living tissues, over that existing in the inanimate environment. This can be very serious, not because the particular radioactive nuclide is different in kind from others, but because concentration through such a mechanism finally leads to a high dose in *rads* to the tissues exposed. Drinking the water might expose one to very little radiation. Eating fish from that water would expose one to 1000 times more cesium-137 radiation, on a weight-for-weight basis.

(d) The role of half-life of the radioactive nuclide

Some of the radioactive substances that occur as a result of nuclear fission in a nuclear reactor are extremely unstable, half of the radionuclide decaying or disintegrating in a matter of seconds. In contrast, radioactive strontium-90 has a half-life of 25 years, radioactive cesium-137 a half-life of 33 years. But what counts, in terms of biological injury to living beings is the number of rads absorbed in a tissue per unit-time, *not* the half-life of the particular radioactive nuclide which is irradiating tissue. Thus, 10 rads from a radionuclide having a half-life of 2 seconds is the same as 10 rads from a radionuclide of 30 years half-life.

However, there is an important difference between the very short-lived radioactive element and the very long-lived one, in terms of potential harm. For the short-lived nuclide, 10 rads delivered in the course of 5 minutes may be *all* the radiation that will ever be received by the tissue, simply because essentially all of the radioactive atoms of that radionuclide have decayed away.

However, for the long-lived radionuclide, not only can it deliver 10 rads in the course of 5 minutes, but it can continue to deliver this amount, or nearly this amount, of radiation every 5 minutes for years and years. This is the essence of the difference in potential hazard for long-lived versus short-lived radionuclides. Rad for rad, however, they are identical.

Another way that half-life of the radionuclide becomes important for biological reasons concerns the *chance* it has to get out of a nuclear power facility in time to do biological harm. If a radionuclide has a half-life measured in seconds, it is clear that if it can be contained for an hour before being released, essentially all of it is gone before reaching any living tissue.

37

On the other hand, radiostrontium-90 or cesium-137 have half-lives of the order of 25-35 years; not only must we worry about them getting out of a nuclear reactor, but we must worry about them for several centuries!

Such radioactive elements must be kept from intersecting with living things, man in particular, at the reactor, during spent fuel transportation and at the fuel reprocessing plant. And finally, the highly radioactive waste must be guarded for something like 500 years. When a radionuclide has a half-life of 30 years, an appreciable amount of radioactivity will persist several hundred years.

(e) Whole body radiation versus partial body radiation

Obviously, irradiation of the entire body with any number of rads is more serious than irradiation of a particular part, some organ for example. It is not easy to state flatly whether irradiation of one specific part of the body is worse than irradiation of some other specific part. In the case of the male and female reproductive organs, however, the damage done during reproductive years guarantees great harm.

Since the gene-containing cells for future generations of humans reside in these organs, irradiation here will cause the genetic (inherited) alterations which can produce mental and physical deformities, and a host of serious diseases in future generations. And since genetic injury is the most serious injury produced by radiation, it is true that irradiation of ovaries and testes is almost as serious as irradiation of the whole body (which, of course, includes the ovaries and testes).

For all the remaining organs, in adults at least, cancer is a major hazard and is almost certainly due, in a specific organ, to irradiation received by that organ. At first thought, it might appear that the organs could be ranked by weight to determine the seriousness of the irradiation for cancer induction. The cancer risk for any specified amount of radiation appears more related to the spontaneous cancer risk for that organ than to its size.

As we have seen, in most cancer induced by radiation, the radiation directly affects the cells which may become cancerous later. However, from certain animal studies, it appears that radiation in one part of the body can result in a lymph cancer developing elsewhere.* This is not common.

(f) Age an important factor

No factor is of greater importance in considering the implications of delivery of radiation to humans than is age. Direct evidence has been provided by Dr. Alice Stewart of Oxford, England that developing embryos are vastly more sensitive to the cancer and leukemia producing effects of radiation than are adults. In fact, a given amount of radiation increases the risk of future cancer or leukemia 50 times more if

---

*"Radiation Injury, Effects, Principles and Perspectives." Arthur C. Upton Ch. III, Radiation Carcinogenesis, p. 70, University of Chicago Press, 1969, Chicago Illinois.

delivered to the embryo during gestation than if delivered to adults. Next to the sensitivity of the fetus in utero are children, and then come adults. Unfortunately, even the sensitivity of adults to cancer production by radiation is 10 to 30 times more than "expert" bodies of scientists thought up until the last few years.

The embryo presents other special problems too. Radiation, received at a time when the various organs are being formed, can cause a whole organ system to be deformed. For example, early radiation can lead to serious brain injury with resultant mental infirmities. This was seen in Hiroshima.

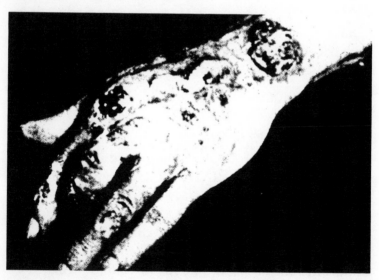

This is the hand of a physician who was exposed to repeated small doses of x-ray radiation for 15 years. The skin cancer appeared several years after his work with x-rays had ceased. Cancer incidence depends on radiation dose.

from Meissner, William A. and Warren, Shields: Neoplasms. In Anderson, W.A.D. editor; Pathology, edition 6, St. Louis, 1971, the C. V. Mosby Co.

Both from the point of view of injuring whole organ systems during pregnancy and that of producing massive increases in the risk of future cancer and leukemia, irradiation early in pregnancy is an extremely serious matter. This would be true for any source of radiation—medical x-rays, nuclear electricity or other activities. Since a woman often does not realize she is pregnant during these critical early stages, it seems extremely unwise for women of child-bearing age to be associated, in any way, with the nuclear power industry or other activities where the chance of radiation exposure is high.

(g) Other radiations: Alpha particles, "Hot" particles, Plutonium

For the broad group of hundreds of radioactive elements produced in nuclear fission, the radiations are beta particles, x-rays, or gamma rays or various combinations. For all these we can focus our attention on the number of *rads* absorbed by an irradiated tissue or organ. This simplification helps.

But one type of radioactivity associated with nuclear electricity deserves special consideration; alpha radioactivity. Alpha particles, electrically charged nuclei of helium atoms, affect material (including human tissues) in their paths so extensively that they travel only short distances before they have expended all their energy. In the process they provide intense radiation to the tissue in their path.

What is more, it appears that for some biological effects, including cancer production, alpha particles

41

are worse per rad absorbed in tissue than any of the radiations discussed above, possibly several times worse.

Much confusion has been generated by some so-called authorities (in AEC) concerning alpha particle radiation. These "authorities" have stated repeatedly that, since alpha particles transfer so much of their energy in such a short distance and are then stopped, it follows that alpha particles are not serious. This assumption is false. It is true that a radionuclide (emitting alpha particles) lodged on the skin cannot irradiate internal tissues, simply because no alpha particles get any deeper than the skin. But they *can* provoke skin cancer.

Much much worse is the *inhalation* of nuclides which emit alpha particles. Once inhaled, the radio-nuclide can be distributed along the lining of the respiratory tract and there irradiate those cells especially prone to develop cancer. Indeed, this is the source of lung cancer induced by radioactive exposure of uranium miners, one of the truly unnecessary tragedies that has already occurred in the nuclear electricity industry.

"Hot" particles are very small dust-like particles that are made up of alpha-emitting substances. One of the prominent ones, plutonium-239, is widely heralded as the "nuclear fuel of the future." Fine particles of pure plutonium-239 oxide (formed when plutonium burns) are very intense sources of alpha particles.

Geesaman and Tamplin have shown that such fine particles, referred to as "hot" particles because of their extremely high alpha particle emission in a localized region may be 10 to 1000 times more effective in producing cancer than would be expected if the same number of rads were delivered in a more diffuse manner to an organ, such as the lung.

It is this "hot" particle problem associated with plutonium-239 that makes the contemplated, future, widespread use of this radionuclide as a fuel in the nuclear-electricity-generation plants such an unmitigated nightmare for mankind. Not only may the hot particles of plutonium oxide be super-cancer producers, but with a half-life for plutonium-239 of 24,000 years, such plutonium oxide can be spread about the earth, re-suspended in air, and produce lung cancers in generations of humans for 100,000 to 200,000 years.

Manufacture of plutonium-239 and its widespread use in nuclear electric power may represent man's most immoral act.

Aside from the alpha-emitting radionuclides and the "hot" particle problem, the vast majority of other radionuclides can be considered to have equivalent effects, provided one considers simply the *rads* delivered to a particular tissue.

These points are stressed because so much confusion has been generated in the public's mind concerning possible special importance of one or another particular radionuclide. The question is commonly asked, "Which radionuclide, associated with

nuclear electricity generation, is to be most feared?"
Aside from the special case of plutonium and the "hot"
particle problem, the answer is any or all the radionuclides in direct proportion to the radiation they
deliver to a particular organ. Important differences in
their *chemical* behavior can determine how, and if,
they will wend their way through the food chain to
man. If the radionuclide gets to deliver its radiation
to man's tissue, only the *rads* delivered count in assessing hazard.

In addition to the rad unit, it has proved convenient to use a unit 1000 times smaller, known as the
millirad. 1000 millirads equals one rad. You may
commonly encounter both the rad and the millirad in
public discussions of radiation hazard. Thus, for
example, the Federal Radiation Council allows Americans to receive an average dose of 0.17 rads per year.
This can also be stated as 170 millirads per year. Both
will frequently be referred to in public writings and
discussions.

Two other units that may be encountered are
the rem and the roentgen. For almost all purposes
involving beta rays, x-rays, and gamma rays, (and
this encompasses all the fission products from
uranium) we can say the rad, the rem, and the
roentgen are roughly equivalent. We shall use only
the rad, or the millirad, in this book. Should the reader
encounter radiation hazards discussed elsewhere in
units of rems or roentgens, there will be no significant
error in considering these units to be the same as rads.

44

CHAPTER 3

How Radiation Produces Disease and Hereditary Alterations

To grasp the significance of the physical harm done to human beings by radiation, it is not necessary to understand exactly what happens in body cells which are irradiated. But we will explain, in several sentences, what is known of these events, in case you may need this information for a debate on nuclear power plants. The terminology may be unfamiliar.

The various kinds of radiation delivered to human cells (from beta rays, x-rays, gamma rays or alpha particles) are commonly referred to as *ionizing* radiation, or radiation which separates or changes into ions. The name is appropriate because the high speed electrons (beta rays) passing through living tissue actually rip negatively charged electrons from atoms, leaving positively charged ions. Such electrons in turn ionize other atoms until finally all the initial energy of the high speed electron is dissipated. Such electrons originate in the nucleus of the unstable, radioactive atom. When emitted, they travel with enormous speeds, some having speeds approaching the speed of light. Many such electrons have enough energy to break 100,000 chemical bonds between atoms.

X-rays and gamma rays, by one or another mechanism, set electrons in motion in tissue. Once this is done, all the events which occur are similar to those produced by an original beta ray. Alpha particles also ionize atoms in their path, setting electrons in motion which cause further ionization. This disruptive action, producing electrically charged ions, is a major, but not the only, way such radiations injure tissues. Many chemical bonds between atoms are shattered in addition to the ionization produced. This is an important additional damage mechanism.

For our purposes, such disruptive actions of ionizing radiation can best be regarded simply as a massive, non-specific *disorganization* or injury of biological cells and tissues. Biological cells are remarkably organized accumulations of chemical substances, arranged into myriad types of sub-structural entities within the cells. The beauty of such organization can only be marveled at when revealed under the high magnifications of such instruments as the electron microscope or the electron scanning microscope. In stark contrast, there is hardly *anything* specific or orderly about the ripping of chemical bonds or of electrons out of atoms. Rather, this represents disorganization and disruption. Perhaps a reasonable analogy would be the effect of jagged pieces of shrapnel passing through tissues. One hardly expects nature's architecture to be improved by the disruptive action of shrapnel *or* ionizing radiation. Instead, we can anticipate varying degrees of damage of the delicate internal cellular architecture.

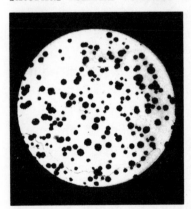 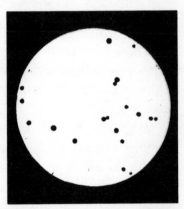

Ionizing radiation can cause reproductive death in human tissue cells. Above are two culture plates showing colonies of human tissue cells. Each was grown from an equal number of parent cells. The parent cells of the colonies on the right were exposed to ionizing radiation, while the parent cells of those on the left were not.

Courtesy of Theodore T. Puck,
Scientific American, April 1960

If the damage is catastrophic, the cell which has experienced the radiation injury dies. If less than that, the cell can go on living, though wounded, for a long time. Not only can wounded cells go on living, they can divide, and reproduce new cells. Unfortunately, these new cells might carry the injury sustained by the irradiated cell from which they originate.

In many body tissues, the loss of a certain number of cells due to radiation damage can be tolerated because remaining, uninjured cells can divide and still maintain the necessary number of functioning tissue cells. Cells that are not injured too badly can carry on their usual function in the body, perhaps at less than optimum performance.

Non-fatal injury to the cells of certain human tissues may be far, far more dangerous to the person than the outright, immediate death of the cell would be. These non-fatal injuries are especially hazardous because, within a period of years, *a single* cell injured in this way has the potential to initiate a cancer or a leukemia.

We still do not know what kind of an injury ionizing radiation induces in the cells that would ultimately lead to a cancer—5, 10, 15, or 20 years later. We do know for certain, that this process does occur. What happens between the initial radiation injury and the ultimate appearance of a cancer or leukemia is still a mystery. But once this process has been initiated by radiation, science knows of no way to stop it.

A wide variety (possibly thousands) of types and degrees of injury to cells may occur from ionizing radiation. Perhaps only one or a few of these may be of the kind which can start a cancer which finally destroys its human host. It is important to realize that one gram of cells (about 1/32 of an ounce) from a human organ contains a *billion* cells, approximately.

So, even if only a very rare type of cell injury (among thousands of possible injuries) can start a cell on the path to cancer, it is still possible for thousands, or hundreds of thousands, of cells to be altered by radiation in a way that will eventually lead to a cancer. Do not forget, *only one* cell with the proper type of radiation-induced disorganization may develop into a fatal cancer. The process in between may be extremely complicated, and many injured cells might not be able to complete all the steps toward production of a cancer. But it takes only one cell to do so.

The period between radiation injury and obvious cancer is quite long in the human. Leukemia, often called blood cancer, takes at least four or five years. Other cancers may take as long as 20 years. The intervening period is silent; the person doesn't realize it is going on. If asked about his health, he would, of course, say, "I feel fine."

Because of the "silence" during which unknown deadly events are occurring, the time between irradiation and appearance of the cancer has been designated as the latency period. So for leukemia, the latency period is some five years; for thyroid cancer, it is approximately thirteen years. For some other cancers, the latency period is still not accurately known although periods of 20 years or more have been suggested.

Once the latency period has passed, a certain type of cancer will continue to appear year after year in a group of humans subjected years before to known ionizing radiation. Acute leukemias due to irradiation continue to appear in apparently undiminished numbers, consistently even 20 years after the Hiroshima-Nagasaki bombing. One form of leukemia, so-called chronic myelogenous leukemia, seems to appear among the exposed population steadily over a period of about 10 years and then to appear less frequently.

For most cancers, we do not know whether they will continue to occur throughout life once the latency period is over. New cases may finally stop appearing after 10 or 20 years in irradiated persons. Most public health officials properly assume that such cancers will continue to appear indefinitely in the irradiated groups. For, the assumption of anything else can lead to grave underestimation of the hazard of radiation.

We must realize that this *major* consequence of radiation injury to cells, namely, cancer or leukemia production, does *not* become evident immediately after irradiation. Sadly, the long delay, or latency period, has proved to be very disarming. The result has been a failure to appreciate and understand the real magnitude of the pernicious effects of ionizing radiation. From radiation and other environmental noxious agents we tend to expect *immediate* effects. If we don't see them, a false sense of security takes over.

The nucleus is generally considered to be the crucial site of cell injury by ionizing radiation. Further, the critical structures injured with the nucleus are the *chromosomes.* In every normal human cell (except for certain stages of sperm and ova cells) there are 46 such chromosomes. These chromosomes are considered by most biologists to carry all, or almost all, the information in the cell, information which directs the cell in all its activities, including growth, cell division, production of a host of biologically-important chemicals such as proteins, and other metabolic activities.

For decades we have known that ionizing radiation can produce microscopically-*visible* injury to these delicate information-bearing chromosomes. Direct breakage of chromosomes into two or more pieces has been observed to occur after irradiation of cells. There is every reason to believe that the chromosomes suffer much additional radiation injury that is not visible under the microscope.

Many authorities suspect that some particular type of chromosome injury, as yet unidentified, is essential if the cell is to go through the sequence of changes that finally convert it into a full-blown cancer cell. Certainly identification of the precise nature of such a chromosomal change represents one of contemporary biology's major challenges. Whatever that chromosome alteration may prove to be, we know that it does occur, all too often, when human cells are exposed to ionizing radiation.

When ionizing radiation interacts with one of the chromosomes, there are two major ways in which the information system of the cell can be permanently altered by radiation. Genes are the units of information within the chromosome. They are composed largely of the chemical known popularly as D.N.A. (deoxyribonucleic acid). Radiation *can* produce a chemical alteration in a part of a single gene, so that the gene functions abnormally thereafter, providing the cell with false directions. When such cells divide, the altered gene may be reproduced in the descendant cells.

If a single gene on a chromosome has been chemically altered, so that it provides new directions, a *point mutation* is said to have occurred. Radiation can also produce a different type of change in the information system of the cell. This change occurs if the chromosome is physically broken. On page 53 is shown a schematic diagram of a human chromosome. It has two arms and a small region between known as the *centromere*. When a cell divides, the centromere leads the way for the chromosome to go to the daughter cell. When radiation breaks off a piece from one of the arms of the chromosome, this piece no longer has a centromere. As a result, it gets lost from the cell on the very next cell division. A single chromosome has hundreds or thousands of genes within it. Thus, the piece of chromosome broken off may have tens or even hundreds of genes in it. Such genes are lost to the daughter cells when their chromosome piece is lost.* Presumably if too many crucial genes are lost thereby, the cell may die.

---

*For a short period of time (measured in hours) a broken piece of chromosome may rejoin its chromosome. Our concern, of course, is with the loss of those pieces which do *not* rejoin their own or some other chromosome in the cell.

Schematically, human chromosomes can be described as follows: each chromosome consists of a centromere and two arms.

One arm may be much longer than the other arm, in certain of the human chromosomes.

arm

centromere

arm

When a cell is preparing for division, the entire chromosome duplicates itself, and the two duplicates are seen side by side.

As the cell completes the division cycle, one member of the pair goes to one daughter cell; the other member goes to the other daughter cell. Any chromosome piece, broken off by radiation, and hence no longer attached to the centromere, is lost to the daughter cell.

To one
daughter cell

To other
daughter cell

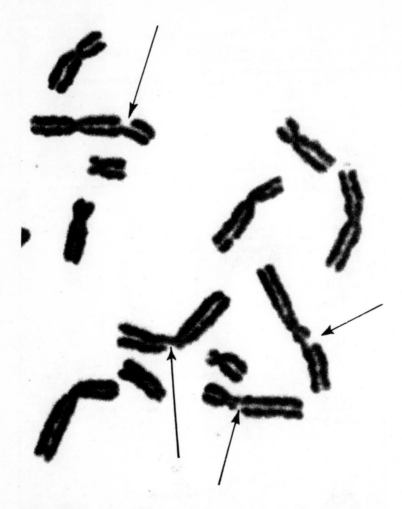

Actual photograph of human chromosomes in a cell that had received gamma ray treatment. Some are intact, others show breaks (indicated by arrows) produced by radiation. The piece which has broken off will be lost when the cell divides. Number of chromosome breaks depends on radiation dose.

With lesser losses, the information alteration is not so grave as to cause the cell's death. But the loss of genes might so imbalance the cellular information in the cell as to cause its ultimate development into a cancer cell.

Loss of a piece of a chromosome and the genes within it is also called a mutation. This loss is appropriately designated as a *deletion,* for we have truly thereby deleted a piece of a chromosome and its genes. So radiation can provoke both major types of mutations, point *mutations* and *deletions.*

If the mutation occurs in a body cell (meaning a cell other than a reproductive cell), the potential result, ultimately, is cancer. The kind of chromosome alteration, or mutation, required is not known. However, leading opinion holds that a single *radiation event* is sufficient to provoke the chromosomal change required in a cell to start it on the path toward being a cancer cell. It is easy to understand from this that as the radiation dose goes up, the risk of future cancer development goes up in direct proportion. This is true because the chance that the "right" kind of single damaging event will occur goes up in direct proportion to the amount of radiation.

New evidence, both for experimental animals and humans, makes it quite certain that the incidence of cancer, after irradiation, goes up in direct proportion to the total amount of radiation received. The particular *kind* of cancer that occurs depends upon which organs received irradiation. Thyroid gland irradiation leads ultimately to thyroid cancer. Mammary gland irradiation leads to breast cancer. Bone marrow irradiation leads to various forms of leukemia.

In each case, the numbers of cancers appearing are expected to go up in direct proportion to the amount of radiation received by the particular organ of the body. Adult nerve cells represent a singular exception. They do not divide, hence, cannot become cancerous. Brain cancer, induced by radiation or occurring spontaneously, is really cancer of special connective tissue cells interspersed among the nerve cells.

Hereditary Alterations

Let us turn now to the effects of radiation-induced mutations in two important remaining cell types, the germinal cells of the testes, source of spermatozoa, the male reproductive cells, and the germinal cells of the ovary, source of ova, the female reproductive cells. Radiation injury to these classes of cells has even more far-reaching consequences than radiation injury which leads other types of cells to leukemia or cancer. Changes in the chromosomes of immature sperm or ova cells may be transmitted to all future generations of humans. The heredity of man, his greatest treasure, is at stake! *Once injured, the chromosomes cannot be repaired by any process known to man.* (Except in the short space of time described above.)

The cells which produce sperm are called spermatogonia. Those which produce ova are called oocytes. Mature spermatozoa have 23 chromosomes. Mature ova have 23 chromosomes. Upon fertilization of the

56

ovum by sperm, we return the number to 46 chromosomes, which characterizes all cells from the fertilized ovum through to the entire adult human.

Injury to the sperm or ova chromosomes while in the testis or ovary, either by point mutation or chromosome deletions (see above), can thus be carried forward into *every* cell of a new human being. Worse yet, since every cell of the new human can carry such a mutation, the sperm or ova of this human can carry them also, so that the original injury persists through successive generations.

We are probably fortunate that some of the mutations have such deleterious effects that the sperm or ova bearing the mutation fail to lead to a fertilized ovum, or if this does occur, the unborn baby is miscarried. But *all too many serious mutations do* permit the development of humans, whose cells bear the mutation, and who suffer serious health consequences as a result.

How serious are the health effects upon new generations of humans carrying mutated genes or altered chromosomes? We are only *beginning* to realize that *it may be possible to tolerate only a very small number of additional mutations of genes or chromosomes as a result of technological poisons if humans are to continue to produce new generations of humans.*

Countless geneticists have repeatedly cautioned society about the danger of allowing *any* increase in the rate at which any type of mutations are introduced into the general population. They know very well that mutations do occur due to natural sources of radiation and to other causes, many of which are not understood to this date. Some who attempt to make light of the hazards of introducing unnecessary mutations are quick to point out that some mutations are beneficial, and indeed they may be. But prevailing genetic opinion indicates that we cannot hope to improve man by increasing his mutation rate. We can, however, count upon doing a great deal of harm, measured in untold human suffering from physical and mental deformities, and a higher incidence of many serious diseases, if we allow mutation rates to increase.

The Nobel Laureate in Genetics, Professor Joshua Lederberg,* recently indicated his grave concern about the implications of increasing the existing mutation rate of our genes, and stated that present radiation standards allow for a 10 percent increase in mutation rate. And he says, "I believe that the present standards for population exposure to radiation should and will (at least de facto) be made more stringent, to about one percent of the spontaneous rate, and that this is also a reasonable standard for the maximum tolerable mutagenic (heredity) effect of any environmental chemical."

Dr. Lederberg is suggesting that *all* forms of influence in our environment which can provoke genetic mutation or chromosome injury be one percent of the

---

*Dr. Joshua Lederberg, Professor of Genetics, Stanford University, Palo Alto, California, Affidavit Sept. 8, 1970 (Docket #3445) before Public Service Board of Vermont.

spontaneous rate, yet he points out the serious situation that we are currently legally permitting 10 percent of the spontaneous rate from radiation *alone*. Let us quote Professor Lederberg on this:

> "A ten percent increase in the existing 'spontaneous' mutation rate is, in effect, the standard that has been adopted as the 'maximum acceptable' level of public exposure to radiation by responsible regulatory bodies."

One wonders how it can be that responsible regulatory bodies would allow ten times more genetic injury to the population from radiation *alone,* when a highly respected geneticist suggests one percent as a maximum for radiation plus chemicals combined. Other geneticists concur.

A multitude of unsatisfactory answers to this question has been provided. One is that we cannot afford to impede technological progress by undue restrictions. Thus, atomic energy programs such as nuclear electricity generation, "must" be beneficial to humans in terms of convenience and comfort, so they must be allowed to pollute the environment with radioactive substances that will ultimately produce genetic changes in man.

A reasonable question: why *must* radioactivity be released at such a high level for atomic energy programs to proceed? This question is never asked, but the answer is, of course, *economics*. It is cheaper to pollute than to take the necessary steps to prevent pollution. Promoters of all technology realize this intuitively and consciously. Hence, they press for the loosest possible standards of pollution or, better yet, no restrictions at all.

And the pressure of such promotional interests is staggering. Generally, all they need to do is mention the magic word "economics," and everyone falls into line. If it is *not economical* to prevent radioactive pollution, then assuredly we must allow the pollution to occur unimpeded. That we may pay an enormous price in the future through deterioration of our genes and chromosomes and, thereby, cause fantastic human misery and suffering, hardly enters this "economic" picture. This is not because the proponents of atomic (or other) technologies are hardhearted, evil individuals, bent upon injury to humans. Far from it.

The *apparent* insensitivity arises from our widespread false definition of the term "economic." We only include short-term considerations in our economic calculations—those concerned with days, weeks, months, or a few years. More ultimate costs to be borne by future society, or future generations, are hard to anticipate (they almost appear "theoretical") and they are routinely avoided in economic considerations.

Another common, but unsatisfactory, answer is given for why we would legally permit enough radiation (and radioactivity contamination) to cause a 10 percent increase in mutation rate. We are already being irradiated, they say, from natural sources (cosmic rays, radioactivity of substances in the earth's crust, carbon-14 produced by cosmic rays) in an amount that can also cause about 10 percent of the spontaneous mutation rate. As this specious argument goes, "we can't do much harm if we do to humans only the equal of what nature is already doing." Fallacious as it is in every respect, this argument seems credible to many among the public, the medical, and the scientific communities.

They *all* fail to realize that natural radiation and the genetic and chromosomal mutations caused thereby *are doing a great deal of harm*. The genetic disorders and deaths caused by natural radiations are no different at all from those caused by man-made radiation. We saw in Chapter II that all these radiations act similarly and the injuries are no different from one source of radiation than from another. All we can say is that, at this moment, we know of no way to turn off the various *natural* sources of radiation. We, therefore, suffer an enormous toll of disease, debility and death as a result of natural radiation. As a minimum element of common sense, we should refrain, except under the most dire circumstances, from adding to this enormous burden of suffering by adding the injury of man-made radiation. The benefits to society should be required to be enormous and *obviously* so before permitting *any* amount of increase in radiation mutations due to man-made sources.

When the argument is raised that natural-radiation-induced mutations cannot be harmful since humans have evolved this far in a "sea of radioactivity," this argument should be countered with several cogent points. First, while we have evolved to our present state in *spite* of radiation, we do have a limited life span and we do have an enormous toll of suffering, disease, and premature death due to genetic disorders. And natural radiation probably accounts for about 5-10 percent of such suffering and disease.

We, societally, are at least humane enough to devote a sizeable share of our funds to medical care and medical research in the endeavor to alleviate the suffering and premature deaths caused by genetic, mutation-induced disease, some 5-10 percent of which is due to natural radiation. We must assuredly think very seriously of having to expend 10 percent more on medical care and consider having the massive increase in disease (genetic) that would go with man-made radiation exacting a toll comparable with or higher than the toll exacted by natural radiation.

Precisely the same foolish argument concerning natural radiation could have been raised concerning poliomyelitis, cholera, typhoid, tuberculosis, yellow fever, diphtheria, and a host of other infectious diseases. It is entirely likely that the organisms causing such diseases have co-existed on earth, with man and other species, for millions of years. Would anyone argue that typhoid fever didn't exist, or yellow fever, or poliomyelitis, or bubonic plague, or diphtheria, or cholera? Hardly! In some areas of the world, life expectancy has not been the classical three score and ten, precisely because diseases caused by such organisms took a heavy toll leading to life expectancies very much shorter than they are today. Who would have listened to the argument that the tubercle bacillus was harmless just because man survived as a species in spite of the ravages of tuberculosis? Who would have argued that same case for the other serious agents of infectious disease?

The situation in regard to radiation injury is actually much worse than the situation in regard to infectious diseases, isn't it? Promoters of nuclear energy are saying, in essence, "Since we already have such-and-such a level of illness from background radiation, it doesn't really matter if we increase this figure to the same level—in other words double it."

Applying this same logic to infectious disease, public health officials would say, "Since we have always had 10,000 cases of malaria in this country, it doesn't matter if we increase the number to 20,000."

Man has, with great ingenuity, searched carefully in his environment for causes of serious disease. Where possible, he has altered the environment, through sanitation, or by immunization procedures, thereby diminishing the enormous toll of infectious disease. What a shame it would have been if man had given up at the start and said poliomyelitis virus has always been with us, man has evolved in spite of it, and, therefore, no consideration need be given to ravages by polio virus.

Precisely how serious are the genetic diseases man suffers from? Extremely serious! This has become increasingly clear to medical authorities from careful studies continuing right up to the present. Before considering the magnitude of the implications of genetically-determined diseases, it is important to point out that new *mutations* of genes and chromosomes are *required* to maintain the occurrence of all diseases that are genetically-determined, with rare exceptions. This is so because ordinarily most mutations introduced into a population render the bearer of the mutation slightly or grossly less likely to bear children than are persons with normal, unmutated genes of that specific type.

Let us consider the most serious genetic (or chromosomal) mutation—which would be one which renders the person bearing the mutation absolutely sterile. In such a case, if a mutation occurs in the ovary or testis of a parent, the offspring may carry the mutation in all of its cells, will suffer the consequences of carrying the mutation, and will fail to reproduce. Thus, this type of mutation will not be propagated in the species beyond the one generation. But it will cause great suffering to the afflicted individual. If, over centuries and centuries the various spontaneous sources of mutation have remained constant, then this particular type of disease will have remained constant, the new cases always arising by mutations in the immediately preceding generation.

If by man-made radiation we increase mutations by 10 percent we can expect an immediate increase of 10 percent (in the very next generation of offspring) in this serious type of disease thereby produced. But because of non-reproduction in such offspring, the effect is not transmitted to additional future generations. So this type of effect of an ill-considered allowance of a 10 percent increase in mutations due to radiation would not continue to persist if we were *then* to discontinue the radiation.

Other genetic mutations do not render the offspring totally sterile but may reduce the average "reproductive fitness" compared to persons with the particular gene in the healthy (unmutated) form. For such mutations introduced by spontaneous sources (radiation or other), there is a *build up* of such mutations throughout the population until the loss of mutated individuals by lesser reproductive fitness just balances the introduction of new mutations of that particular type. The human species must have reached equilibrium in this sense, since if spontaneous mutations have been going on for millenia, by now the production rate and loss rate are equal. Disease due to such mutated genes is occurring in *every* generation.

If, now, we increase the mutation rate by 10 percent due to man-made radiation and keep on doing this generation after generation, what will happen? Since many people already have that mutation from the previously established equilibrium, we will be adding to *that* number those due to the increased mutation rate, until after *some* number of generations (not precisely known, for it depends upon reproductive fitness) the loss of individuals by diminished fertility will balance the increment produced by the radiation. We will have a new equilibrium BUT now there will finally be 10 percent more individuals in the population bearing the mutation and, hence, there will be 10 percent more of the biological damage produced per generation by this particular defective gene or chromosome. The cost in health per generation can be *much more serious* than what would be expected just from the 10 percent increase in persons bearing the mutation. We shall see precisely how this can occur as we turn attention to the kinds of diseases caused by defective genes.

What Kinds of Genetic Diseases?

In recent years in medicine, our horizon has broadened considerably concerning the implications of genetics and mutation for human disease. In the past, genetic diseases were considered to be a relative rarity among the causes of disability and death. Now we realize that this rarity was an illusion, which led to a *grave* underestimation of the role of genetic mutations in human diseases. Today we recognize that a large proportion of all human afflictions are at least partially determined by heredity, and hence related to genetic mutations. Numerous authorities and authoritative bodies consider that the developing evidence may finally indicate that *most, if not all,* human disease has a genetic component.

The United Nations Scientific Committee on the Effects of Atomic Radiation states:*

---

*Report of the United Nations Scientific Committee on the Effects of Atomic Radiation. General Assembly. Official Records: Seventeenth Session. Supplement No. 16 (A/5216) Chapter IV, "Hereditary Effects," paragraph 56, page 19.

"It is generally accepted that there is a genetic component in much, if not all, illness. This component is frequently too small to be detected; in other instances, the evidence for its presence is unequivocal. Nevertheless, the role of genetic factors in the health of human populations has not in the past been considered seriously in vital and health statistics. As a consequence, data on the prevalence of hereditary diseases and defects are now largely restricted to that collected by geneticists for special purposes in limited populations from a small number of countries. An assessment of the hereditary defects and diseases with which a population is afflicted does not necessarily provide a measure of the imposed burden of suffering and hardship on the individual, the family or society."

Professor Lederberg* has recently stated the following:

"We can calculate that at least 25 percent of our health care burden is of genetic origin. This figure is a very conservative estimate in view of the genetic component of such griefs as schizophrenia, diabetes, and atherosclerosis, mental retardation, early senility, and many congenital malformations. In fact, the genetic factor in disease is bound to increase to an even larger proportion, for as we deal with infectious disease and other environmental insults, the genetic legacy of the species will compete only with traumatic accidents as the major factor in health."

Professor Lederberg has stated the problem succinctly and well. In the earlier days of medicine our techniques of sorting out genetically-determined diseases were cruder and tended only to find the diseases that had a simple so-called Mendelian form of inheri-

---

*"Government Is Most Dangerous of Genetic Engineers," Joshua Lederberg, *The Washington Post,* Sunday, July 19, 1970.

67

tance. These are diseases which could be referred to as single-gene diseases. The inheritance patterns expected were known, and hence the genetic basis for the diseases ascertained, relatively easily, by studies of the occurrence of the disease in families and their ancestors.

Among the classical cases of such diseases are the well-known phenylketonuria, galactosemia, cystic fibrosis, sicklecell anemia, hemophilia, and others. However, altogether such diseases, numerous as they are, only accounted for less than one percent of deaths. This is very serious, but still is small compared to the now greatly expanded list of genetically-determined diseases, the now well-known multigene diseases.

For many years, medical experts realized that a host of the more common and serious diseases of man had a familial pattern, but not one as readily ascertainable as was the case for the single-gene diseases listed above. Dr. C. O. Carter, in a recent compilation of the evidence,* has shown that a whole group of important human diseases are indeed genetically-determined, but it appears that these diseases are determined by the interaction of more genes than one, and that this is complicated by further interactions with environmental factors.

As a result of such work, we now are forced to consider not only the rarities like hemophilia as geneti-

---

*"Multifactorial Genetic Disease" by C. O. Carter, *Hospital Practice* Vol. 5, pp 45-59, May 1970. (Dr. Carter is Director, The British Medical Research Council's Clinical Genetics Unit.)

cally determined diseases, but also diabetes mellitus, atherosclerosis (the major form of hardening of the arteries), schizophrenia, and rheumatoid arthritis *all* as being genetically determined diseases. Hence, they are *all* subject to increase in occurrence as a result of increase in genetic mutation rates by radiation or any other mutagenic influences.

How do such diseases, added to the genetic list, lead Professor Lederberg to say a conservative 25 percent of all diseases are genetic, or lead others to say possibly all diseases (aside from trauma) may have a genetic component? Let us focus our attention on the disorder known as atherosclerosis. This disorder underlies most cases of the most serious form of heart disease in the USA, namely, coronary heart disease. It is coronary heart disease that accounts for the great majority of "heart attacks." And coronary heart disease kills more than twice as many Americans, prematurely, as all forms of cancer plus leukemia combined!

What is more, atherosclerosis not only affects the arteries of the heart, but also those of the brain, many internal organs, and the legs. The total disability and death from atherosclerosis are really not fully realized at all, for as a complicating factor in other diseases, its role may have been underestimated—and underestimated seriously. The fact that atherosclerosis and coronary heart disease must now be regarded as genetic in origin, really means that over 50 percent of all disease, at least, is genetic. The implications of genetic mutations are thereby rendered grossly more serious than realized previously, when only single-gene diseases like hemophilia were considered as the genetic disorders of man.

It was stated before that a 10 percent increase in genetic mutation rate would ultimately lead to 10 percent more of the biological damage produced per generation by this particular defective gene or chromosome. The cost in health per generation *can* exceed the 10 percent increase in biological damage. Let us consider atherosclerosis again. While we know that more atherosclerosis will result in a higher frequency of heart attacks, we do not know the precise relationship between degree of atherosclerosis in the arteries of the heart and the occurrence rate of heart attacks. Indeed, the available evidence on this subject suggests that the risk of a premature heart attack may rise *much more steeply* than simply in proportion to the degree of atherosclerosis. It may well be that an increase of 10 percent in the *average* degree of coronary artery atherosclerosis may lead to a 50 percent increase in the frequency of heart attacks. We simply don't know this relationship well enough. Similarly, atherosclerosis of the arteries of the brain underlies a fair proportion of "strokes," or cerebrovascular accidents. Again, whether a 10 percent increase in average degree of atherosclerosis of the cerebral arteries will increase strokes by 10 percent, 20 percent, or 50 percent is just not known.

As a result, while we can anticipate that a 10 percent increase in mutation rate will ultimately increase the biological damage resulting in major diseases by 10 percent, it is also quite possible that the increased disease *incidence* may exceed this 10 percent increase in damage (already of grave consequence) by quite a lot. The consequences of genetic mutation, as a result of the new medical concepts of the important role of genetic factors in health and disease, are indeed far, far more serious than were realized 10 short years ago.

Incidentally, many of the standards for so-called "allowable" doses of radiation to the public for atomic energy programs such as nuclear electricity generation were set before the new implications of human genetic diseases were appreciated! This fact alone requires a total re-evaluation of atomic energy programs, nuclear electricity generation among them.

CHAPTER 4

Is Any Radiation "Safe"?

When the industrial application of a technology has a poisonous environmental by-product, clearly a primary requirement is to understand the consequent public health hazard. Past experience shows that, at any point in medical history, we are unlikely to know enough to assess such hazards with accuracy. But the consequences of underestimating these hazards, especially genetic hazards, can be so grave that we must be extremely conservative in assessing them.

An error on the side of conservatism in estimating a danger can be, at worst, a delaying nuisance for the promoters of the technology. An error on the side of optimism, leading to some underestimation of the true hazard, can be extremely costly to the human species. We can always *later* allow more exposure to a poison, such as radioactivity, if we learn that it can be tolerated. We *cannot* undo genetic and chromosomal damage from overdoses of poison already consumed.

For nuclear electricity generation, the by-product poison is radioactivity (or radiation itself). Any of the hundreds of radioactive substances produced in the course of all phases of nuclear electricity generation can be harmful to man, from uranium mining through to disposal of astronomical quantities of radioactive wastes. It doesn't matter whether the radiation is external to the body or provided by one or more radioactive compounds that have gained access to the body through air, food, or water. What counts, for any particular organ, is the total absorption of radiation energy, which is measured in rads or millirads (1000 millirads = 1 rad).

The only possible way to set a truly *safe* standard, — a definite number of rads or millirads assigned to a particular tissue or organ — would be to know beyond any reasonable doubt that within that amount no biological effect will occur. We can state unequivocally, and without fear of contradiction, that no one has ever produced evidence that *any* specific amount of radiation will be without harm. Indeed, quite the opposite appears to be the case.

All the evidence, both from experimental animals and from humans, leads us to expect that even the *smallest* quantities of ionizing radiation produce harm, both to this generation of humans and future generations. Furthermore, it appears that progressively greater harm accrues in direct proportion to the amount of radiation received by the various body tissues and organs.

It came as a great shock to us, in the course of our study of radiation hazards to man, that nuclear electricity generation has been developed under the false illusion that there exists some safe amount of radiation. This unsupportable concept is surely one of the gravest condemnations of nuclear electricity generation. Obviously any engineering development proceeding under an illusion of a wide margin of safety is fraught with serious danger.

What is more, the false illusion of a safe amount of radiation has pervaded all the highest circles concerned with the development and promotion of nuclear electric power. The Congress, the nuclear manufacturing industry, and the electric utility industry have all been led to believe that some safe amount of radiation does indeed exist. They were hoping to develop this industry with exposures below this limit — a limit we now know is anything but safe.

Before describing the widely pervasive nature of this serious misunderstanding of the radiation hazard problem at such top levels in industry and government, it is important to establish carefully that we put the integrity, sincerity, and motives of *no one* into question. Undoubtedly, the scientists, the engineers, and the power executives involved, as well as the Congressmen, were simply misled in their belief that some safe amount of radiation truly exists. It was the result of some inadequate observations involving persons exposed to radium salts industrially. Numerous reputable scientists had long discounted these inadequate observations. All of the national and international standard-setting bodies had also refused to accept this inadequate evidence of a supposedly safe amount of radiation.

A Multi-Billion Dollar Industry on a Dangerous Premise

How, under such circumstances, is it even conceivable that so many important industrial and governmental leaders were so totally and seriously misled, misled to the point of launching a multi-billion dollar industry based upon a dangerously false premise?

One agency must bear the responsibility for this wrong impression — the United States Atomic Energy Commission. Probably there was no willful wrongdoing. But the Atomic Energy Commission, burdened by Congress with the impossible dual role of promoter of atomic energy and protector of the public from radiation, has historically suffered from false optimism.

This is not only true, as we shall see, for radiation hazard issues, but also concerning the economics and the capabilities of the nuclear industry. The AEC, of all the agencies we know of, was the only one to back the idea that a safe "threshold" amount of radiation existed. Apparently it was able to convey this belief, not only to the Joint Committee on Atomic Energy, but to the nuclear and electric industries as well.

This untenable concept, that there does exist some dose in rads, or millirads, innocuous to humans, is a grave danger to the public. Indeed, it would be difficult to conceive of a more serious situation in a burgeoning industry than to have such appalling misinformation rife among corporate executives and their leading technologists.

It is our good fortune that the misunderstanding has been uncovered at last. It is now possible to halt unsafe growth of the nuclear electricity industry before the errors of the past are compounded.

We believe the reader will be, and should be, interested in how we became aware of the widespread lack of understanding of the radiation hazards which characterizes the highest circles of atomic energy. Our realization arose from a direct encounter with the very top of the U.S. pyramid of atomic energy, namely, the Joint Committee on Atomic Energy of the U.S. Congress.

We were assigned to evaluate the hazards of atomic radiation by the U.S. Atomic Energy Commission in 1963. It was our job to assess the cost in human disease and death for all sorts of proposed and on-going nuclear energy programs, including nuclear electricity.

In October, 1969, we were prepared to present our results on the expected cancer and leukemia deaths for human exposure to various amounts of radiation. We could have presented these results as the number of

77

cancers or leukemias per year per rad or per millirad of exposure. The most useful basis appeared to us to express this risk for that amount of radiation, namely, 0.17 rads (or 170 millirads), which is the currently *legal average* dose that peaceful atomic energy programs are permitted to deliver to the U.S. Population.

This dosage, 0.17 rads per year for average exposure, is a "guideline" of allowable dosage set by the Federal Radiation Council, an organization established in 1959 by President Eisenhower. This guideline in no way suggests that *everyone* should or does receive 0.17 rads per year. Rather, it states that *no* individual shall receive *more* than 0.5 rads per year nor shall the average dose exceed 0.17 rads.

Thus, any combination of atomic energy programs could go forward legally, provided these criteria were met. It is conceivable that atomic energy programs might be irregularly distributed throughout the country, with the average exposure in some regions at 0.34 rads and in other regions at 0 rads. If these two regions were of equal size (in population) the overall national average would be 0.17 rads, which is perfectly legal under the guidelines. And let it be underscored that no one could be taken to task for allowing radiation exposure of the public up to the full limit of this *allowable* dose.

Our calculations of cancer hazard were presented to an eminent scientific society, the Institute for Electrical and Electronic Engineers in San Francisco on October 29, 1969. The prediction follows:

> "If the average exposure of the U.S. Population were to reach the allowable 0.17 rads per year average, there would, in time, be an excess of 32,000 cases of fatal cancer plus leukemia per year, and this would occur year after year."

When we presented this estimate, we anticipated no opposition whatsoever to our scientific findings.

Our work showed that previous estimates (aside from early correct estimates by Professor Linus Pauling) were 10 to 20 times too low. The new evidence, on radiation-induced human cancer-plus-leukemia, from Japan, from Great Britain and from Nova Scotia, were now all telling us one story — radiation is a greater factor in cancer-leukemia than had been previously realized.

New evidence was in; we were simply taking it into account. We expected the nuclear electricity industry and the U.S. Atomic Energy Commission to welcome our report on the cancer + leukemia risk — especially since the findings were being made available *before* a massive burgeoning of the nuclear electricity industry.

At that time (October 1969), we had not given any special thought to the nuclear electricity industry per se. In fact, in our preoccupation with a careful analysis of the hazard per unit of radiation received by people, we had thought nuclear electricity one of the most innocuous of atomic energy programs, a view we have now had to alter radically. What surprised us beyond belief was that from all over the country our colleagues in various aspects of nuclear energy, particularly nuclear electricity, expressed their shock and disbelief that such a massive cancer-plus-leukemia risk could conceivably accompany exposure at the "allowable" Federal Radiation Guideline.

Many of the people, in nuclear electricity work, simply expressed their disbelief that a Federal Agency would ever set a guideline that could be associated with such an enormous hazard. They all had been under the illusion that the hazard at the Guideline radiation level must be *zero*, or at least so very low as to be negligible. One after another, officials of the nuclear electricity industry expressed their opinion that surely *something* must be wrong with our estimates, although none of them could muster an iota of evidence as to what it could be.

Only then were we alerted to the alarming state of affairs — a whole new industry, nuclear electricity, was growing up in the country with all of its experts totally unaware of the true hazards associated with it. We are not speaking about the kind of natural defensiveness a mother shows on hearing unkind remarks concerning her child. This type of defensive reaction on the part of some nuclear electricity spokesmen was quite understandable.

But the sincere lack of realization by the nuclear electricity industry, that there is no proof of safety for *any* amount of radiation, was really disturbing. Our alarm reached its most serious proportions a few weeks later when we came to understand that even Congressman Chet Holifield, *Chairman* of the Joint Committee on Atomic Energy of the U.S. Congress was totally misinformed concerning radiation hazard. And the Joint Committee on Atomic Energy is the chief source of promotion of the nuclear electricity industry!

Congressman Holifield called us in shortly after our presentation of the cancer hazard and expressed his dismay at our pessimistic projections. He told us he had been *assured* that the hazard level was truly 100 times higher than the level at which the Federal Radiation Council Guidelines had been set. He had *assumed,* therefore, that the "allowable" dosage couldn't possibly be associated with the production of *any* cancers or leukemias.

How is it even conceivable that the top man in this country's nuclear energy development could be so totally and brutally misled and misinformed? There is only one answer — the U.S. Atomic Energy Commission had failed utterly to provide Chairman Holifield with the *real* information concerning radiation hazards.

The AEC surely realized that all the responsible national and international radiation protection bodies were on record rejecting the idea of any safe amount of radiation (for cancer production.) Can it be that the dual role of promoter *and* protector made it difficult for AEC to inform Congressman Holifield properly? As the Chairman of the Joint Committee he is the *one* American in government who needed to know the truth concerning radiation hazards.

We can ill afford, on such desperately important

81

issues, to have key governmental officials so totally and hopelessly confused. No wonder this ignorance and confusion spread to the leaders of the nuclear electricity industry. These executives and engineers assumed, quite reasonably, that the Chairman of the Joint Committee on Atomic Energy must know the facts.

We readily discovered how Chairman Holifield had been misled, and we so informed him in the following letter of December 1, 1969.

> UNIVERSITY OF CALIFORNIA,
> LAWRENCE RADIATION LABORATORY,
> Livermore, Calif., December 1, 1969.

HON. CHET HOLIFIELD,
Chairman, Joint Committee on Atomic Energy,
U. S. Capitol, Washington, D.C.

DEAR CONGRESSMAN HOLIFIELD:

Both of us were deeply honored by the opportunity of some two hours of frank and substantive discussion with you and your colleagues last week. Especially is this so because both of us are intense admirers of the devoted and untiring efforts of the Joint Committee on Atomic Energy to bring to light all the true facts concerning radiation hazards. The various Hearings you have held are unequalled as a monumental contribution to the public welfare and health.

In our discussion you asked us a very specific question, "How can you tell us there is a potential hazard at certain dosages when we have been assured that the hazard level is approximately 100-fold higher?"

We answered, "Congressman Holifield, we believe you have been misinformed."

We know that you needed more answer than that. Based upon the evidence and calculations, we knew that what we were saying had to be true, but we did not know *how* it had come about that deep *mis*-information had come to the Joint Committee. We resolved, therefore, to go right home and find out how this had, indeed, come about. After careful study of many of the Hearings of the Joint Committee on Atomic Energy, we believe we have complete understanding of the *specific nature of the misinformation.*

It is the purpose of this letter and the attachments to explain all of this to you. And we are prepared to defend our analysis of this situation in any format the Joint Committee would find helpful. We believe, however, our analysis will speak for itself.

Specifically, we refer to the Radium Dial Painter studies reported to you by Dr. Robert Hasterlik at the Hearings (87th Congress, Part 1, p. 325) and by Dr. Robley Evans at the Hearings (90th Congress, Part 1, p. 265).

Dr. Hasterlik interpreted his findings correctly when questioned by Congressman Price. Dr. Evans, in our opinion, grossly *misinterpreted* his own data, but undoubtedly with total sincerity of purpose.

Our analysis attached shows both sets of data consistent with each other. In *striking contrast* with Dr. Evans' claim that the data indicate a *threshold* of radiation below which cancer doesn't occur, our analysis indicates *nothing of the sort.*

1. Neither the Hasterlik data nor the Evans data *can even remotely* be construed to suggest any safe "threshold" below which cancer doesn't occur.

2. The data from both researchers are perfectly consistent with cancer production right down to very low doses, and this could very well be a linear relationship over much of the entire dose range from low doses upward.

We are both dismayed that the Editorial Board of the "British Journal of Radiology" and the Editorial

Board of "Health Physics" did not catch the indefensible claim of Dr. Evans that a threshold exists.

Worse yet, we are dismayed, indeed, by Dr. Evans
statement that his "proof" of a threshold is the *cornerstone* of all radiation protection standards. If this be
true, then, there is little wonder that the cornerstone
of radiation protection standards is made of quick-
silver.

We believe, after careful study of this particular
fiasco, you may be more understanding of our total
lack of confidence in the underlying basis for existing
radiation standards. However, we are certain everyone concerned in informing you was well-intentioned.

Since we know this information will be of great
interest to the AEC, we feel you will approve of our
sending copies of this letter and the enclosures to
Chairman Seaborg and Dr. John Totter.

Assuring you of our deepest commitment to constructive assistance to you in your gravely important
responsibilities, we are

> Sincerely yours,
> JOHN W. GOFMAN,
> ARTHUR R. TAMPLIN.

Faith Is Not Justified

There are, in addition, further mechanisms which
operated to confuse the Congress, the electric utility
industry, and, eventually, the public of the United
States. It is commonplace to find that people in industry, the Congress and the public-at-large, have an
almost mystical faith in governmental regulatory
agencies. That faith led them to believe that no governmental agency would ever dream of setting a "permissible" dose of radiation that could add one new cancer
case for every 10 already occurring — as the Federal

Radiation Council apparently had. By examining this situation closely, we come to understand that the erroneous faith comes from "failing to appreciate what was written in fine print."

The story begins with the emotional and vehement controversies over the hazard of radioactive fallout from nuclear weapons testing of the 1950's.

President Eisenhower, duly disturbed over these controversies, and hopeful of ameliorating this situation in the atomic energy field, established a new agency, the Federal Radiation Council. This agency was to review the evidence concerning radiation hazards, and to provide guidance for the nation concerning what exposure might be regarded as permissible in association with further developments in the atomic energy field. It is this same Federal Radiation Council which finally set 0.17 rads as a legally-permissible, average, annual radiation exposure for the U.S. population, associated with "peaceful" uses of the atom, such as nuclear electricity generation.

And here is where the fine print must be carefully examined! In setting this permissible limit of 0.17 rads per year, the Federal Radiation Council did *not* say that this radiation dose was expected to be safe. Far from it! What they said, in effect, was that they *hoped* the benefits to be received from peaceful uses of the atom would outweigh the *risks* associated with their permissible doses to the population.

A reasonable person, having studied this "fine print" qualification might want to know how the

Federal Radiation Council had performed the benefit evaluation and the risk evaluation, both so essential in reaching the hopeful conclusion that the benefits outweigh the risk? But, alas, nowhere is there any evidence that this critical part of the task was even *attempted* by the Federal Radiation Council.

The public and industry, not realizing the fine print qualification, assumed that "permissible" meant "safe." Little did they, the public and industry, know that "permissible" doses could ultimately translate into disaster.

The Federal Radiation Council can deny all culpability — for all they were suggesting was the *hope* that the benefits would outweigh such risks. The Atomic Energy Commission cannot escape justifiable criticism and condemnation, for they have endeavored, and still endeavor, to create the impression that "permissible" radiation exposure means "safe" radiation exposure.

As a result, utility company officials make countless public statements, and written pronouncements to the effect that 0.17 rads would be without harm to the individuals receiving this amount of radiation. Witness a typical statement by Mr. Frederick Draeger of the Pacific Gas and Electric Company:*

"There is no evidence that 170 millirads is harmful and any new plant will actually emit only an infinitesimal fraction of that amount."

---

*Quoted in "Nuclear Hazard In Santa Cruz" by Harold Gilliam, San Francisco Chronicle, Sunday, June 28, 1970.

Apparently, Mr. Draeger hasn't the slightest comprehension of what his statement "no evidence" really means. "No evidence" here means no one has even looked!

Considering their sophistication and the high degree of responsibility they exercise, the brainwashing of the corporate executives and engineers of the largest single industry in the U.S., the power industry, is hard to believe. Certainly it deserves the No. 1 position among the modern marvels of public relations efforts.

Faced with the fact that a multi-billion dollar industry has been led down a path of potential economic and public disaster, the Atomic Energy Commission could, of course, also retreat from responsibility by saying, "We too hoped the benefits would outweigh the risks." After all, the AEC hadn't specifically and officially claimed absolute safety. But AEC officials have repeatedly obscured the knowledge of potential hazards! The unfortunate result, even if unintended, is a widespread, false impression of safety in the standards set for nuclear electricity generation and other atomic energy programs.

In their shabbiest efforts at self-exoneration, both the Federal Radiation and AEC officials have stated that they were not *advising* the irradiation of all members of the population up to the "permissible" limit. Amen! If we must all be grateful for small favors, this certainly must be one.

Recently HEW Assistant Secretary Roger Egeberg

stated in Congressional hearings (August 5, 1970):

> "The FRC position at the present time can briefly be summarized as follows:
>
> 1. We continue to advocate the basic premise that the FRC guides must not be construed as an 'allowed' dose which could result in every person in the United States eventually being exposed up to the allowed level."

This remarkable statement of the FRC position would be ludicrous if it didn't deal with such a deadly serious threat to the future health of the entire U.S. population. If the FRC doesn't want the U.S. population exposed to it, what ever led them to set such a guideline as the *allowable* exposure? A reasonable question suggests itself: "If you want the exposure to be kept at some *low* level, why not set the *allowable* dose there?"

The Federal Radiation Council antics do, in many respects, compete with those of the Atomic Energy Commission, possibly because the AEC is represented on the Federal Radiation Council. For example, the Federal Radiation Council has stated it is "inadvisable" to accept any amount of radiation without good reason. But where did the FRC *ever* present any "good reasons" for allowing 0.17 rads as the average U.S. population exposure?

Considering the magnitude of the cancer, leukemia, and genetic fatalities to be expected from such "guideline" allowable exposure, an incensed public should very well demand, from the Federal Radiation Council, explicit good reasons for allowing 0.17 rads per year.

A more farcical antic is displayed in some recent testimony of Dr. Paul Tompkins, Executive Director of the Federal Radiation Council—testimony published in the record of the Hearings of the Joint Committee on Atomic Energy:

Quote of exchange between Dr. Paul Tompkins (of FRC) and Senator Pastore (JCAE):

Dr. Tompkins: "No, sir. It is an exceedingly difficult problem because when dealing with radiation hazards and accepting the no thresholds, the classic approach of setting "a safe level" is denied to us.

Consequently we have to look at the magnitude of the risk on the one side, and the operational requirements, the desirability of the activities, and so forth on the other, which means there has to be a consensus among many conflicting viewpoints and getting such a consensus is often very difficult."

Chairman Pastore: "May I ask you a question on that point? Reverting back to your statement on page 2. Does the Federal Radiation Council concern itself with the economic aspects of this problem? Does that make a difference to you as to the health measures that are to be recommended?"

Dr. Tompkins: "It depends upon how you define economics."

and, additionally

Senator Pastore: ". . . I agree with that statement, of course, but I was wondering within the province and the purview and the functions of the Federal Radiation Council what are the guidelines that you take into account in reaching a decision?"

Dr. Tompkins: "Primarily the health consideration is, of course, the overriding factor."

Chairman Pastore "There again you are leaving room for something else. Why shouldn't it be the only factor, from your point of view?"

> *Dr. Tompkins:* "Well, if one can provide a statement as to how much individual risk might be acceptable for certain activities, then I think that would be the only consideration needed."

And separately from a statement of Dr. Tompkins in those same Hearings (p 34):

> "The primary objective of the FRC is to make recommendations which represent a reasonable balance between biological risk and the impact on uranium mining."*

What "operational" requirement of any aspect of the nuclear energy program is more important than the health and safety of the people of the United States? Precisely *who* decides upon "operational requirements" for nuclear electric power production that might sacrifice thousands or tens of thousands of *additional* human lives to disease annually.

The FRC *always* has expressed its grave concern about "operational" requirements, meaning convenience of governmental or industrial atomic pollutors. The FRC always has expressed its grave concern about the cost in *dollars* of protecting humans from senseless radiation. We have yet to see the FRC express grave concern about extra degenerative and genetic diseases from its own "guideline" radiation doses.

One further group of totally irresponsible statements emanating from nuclear power proponents deserves careful study here. These statements are not only misleading, but they can be construed in one of

---

Reference: "Radiation Exposure of Uranium Miners." Hearings of the Joint Committee on Atomic Energy, 90th Congress, 1st Session. May-August, 1967. Part 1.

only two ways. Either those who make the statements don't understand what they are saying, or they deliberately intend to deceive the public.

Recently Dr. Theos Thompson, one of five U.S. Atomic Energy Commissioners, said:

> ". . . Obviously this is a very small amount of radiation compared with the levels which mankind has been receiving through all of the ages. To date, in spite of many careful studies, no one has been able to detect any effect from these low levels of radiation and it is unlikely that studies of literally millions of cases would show any such effects."*

First, we challenge the AEC to produce a *single* careful study on this issue.

Furthermore, at first reading the unsuspecting public will again understand this statement to mean that low doses of radiation will produce no effect upon humans. Did Dr. Thompson really want to say that? Why did he make the statement at all?

"No Effect Observed"

First, the AEC can retreat, upon challenge, to a position that all that was meant was "no effect *observed.*" AEC is not claiming "no effect *occurs.*" The public would have every right to be outraged by this shallow defense of an indefensible statement. But the "no effect observed" statement is not unique to any one spokesman of the AEC. Governmental and indus-

---

*Reference: "Power Technology and The Future" by Commissioner Theos Thompson (USAEC). Presented at "Briefing Conference for State and Local Government Officials on Nuclear Development," Columbia, South Carolina, May 21, 1970.

trial atomic energy hucksters seem to adore this statement.

Suppose there were 1,000 persons in an auditorium and suddenly the lights were extinguished. During the period of ensuing darkness in the auditorium, suppose a man is stabbed to death. When the lights go on again, it is perfectly appropriate (in Dr. Thompson's framework) to state that no murder was *observed* ("no event observed"). Yet there is a result for certain — in the form of a murdered man!

What does this analogy teach us? Simply if we do not *look,* or if it is too dark to *see,* then no event can be observed — no matter what disastrous result has *occurred.* We have every right to be shocked that such devious, non-reasoning pronouncements are typical of nuclear electricity promoters. Indeed, such pronouncements are so characteristic of them that the sheer repetition of such nonsense probably leads even them to believe what they say makes sense.

What one really must ask the soothsayers of nuclear energy, who appear to have no understanding of public health principles, is, "Have you ever looked properly?" Or, "Were the lights on when you looked?" Their answers we fear, will be meaningless or non-existent.

They may point out to us, for example, that people have received ionizing radiation from medical x-rays or at their work and "no effect has been observed." What these so-called atomic authorities *mean* is that after exposure to 5 rads, for example, people don't seem to die immediately.

This has absolutely nothing to do with whether deadly effects are occurring. For what worries us, and *should* worry every American concerning the ill-conceived, burgeoning nuclear electricity industry is totally different. We don't expect all exposed persons will die immediately or next week. It is cancer or leukemia five or ten years later, and genetic diseases in many future generations of humans that concern us. Major human tragedy can be occurring and yet, with closed eyes, "no effect is observed."

Over and over again the public is treated to "no-effect observed" pronouncements by AEC officials, such as Commissioner Thompson and Commissioner Larson, when it is quite clear that no meaningful study was ever made. No such studies exist.

On the other hand, Dr. Alice Stewart (*Lancet*, June 6, 1970) has produced solid evidence that 250-350 millirads delivered to embryos (1 x-ray film) during gestation produces about a 25% increase in the subsequent occurrence of childhood cancers and leukemias. Faced with such evidence we wonder very seriously whether AEC Commissioners would really continue to make the deceptive, irresponsible statements concerning "no effect observed."

We have pointed out the treacherous nature of the statement "no effect observed" used by atomic energy proponents to justify allowing population exposure to radiation. Were this an isolated example, now past, we could realize this and forget it. But what about tomorrow?

We must ask ourselves, "Would it be possible for a major public health calamity to occur, due to a by-product poison, and go unappreciated until it is too late?" The answer is, it could readily occur if officials continue their application of "no effect observed." The reason is that few of these officials appreciate what a major public health calamity *is!*

Many persons think of poisons in a "one-to-one" sense. They expect a very high proportion — one out of two or one out of ten people — to show a serious effect. And beyond this, the expectation is that the effect will occur soon (hours, days, or weeks) after the exposure to the poison. If we apply such expectations to potential environmental injury, the human species would surely be doomed. Disastrous effects can occur and be far more subtle than this.

How disastrous? The combined toll of misery and death due to *all* forms of cancer plus leukemia would certainly be regarded by every American as a major human tragedy. We know this is true, for Americans consider a reduction in the burden of suffering from these diseases as a major priority goal of generously-supported cancer research. Moreover, research is generously supported for the much more limited, and modest, goal of an added 6 months to a few years of life for the victims of cancer or leukemia.

In the United States some 320,000 people die annually from all forms of cancer plus leukemia combined. The number appears large. But with 200-million persons in the U.S.A., the fatal toll of cancer is *one person annually out of every 600 people.* It may come as a surprise that a major human disease, cancer, strikes "only" one out of 600 each year. Intuitively one expects "a major killer" to strike many more persons per year.

This same type of thinking makes it easy to overlook the introduction of a *major killing* disease of cancer's magnitude by a cavalier approach to environmental questions.

Such a disaster can be introduced easily and unobtrusively because of two fundamental errors in public health thinking:

(a) We tend to look for "immediate" effects of poisons.

(b) We forget what careful studies are required to show that 1 out of 600 die per year of a disease.

And this is the kind of erroneous public-health thinking that encourages the specious statements of technology promoters that "no effect has been observed."

The kind of serious poisoning produced by radiation has already been described. We know that cancer and leukemia *begin* to occur 5 or more years *after* the radiation has been received. Thus, if radiation were the environmental insult under study, one could examine a group of exposed persons after one, two, three, or four years, and reach the massively erroneous conclusion that the radiation had done no harm. This is what happens if scientists look for *immediate* effects when the real hazard is delayed. A self-evident truth? We must remember that one of the greatest public health errors of judgment was made in precisely this way in the field of radiation protection—and made by highly competent scientists!

Leukemia appears in radiation-exposed persons approximately five years after exposure, whereas most other cancers take 10 or more years to occur. Therefore, a group of humans exposed to ionizing radiation would show only leukemia five years later, simply because all of the other cancers had not yet occurred. The expert bodies of scientists studying radiation hazard for humans fell into this specific trap and as a result, they seriously underestimated the cancer hazard.

What is even worse, the error has been even further compounded. Knowing that some forms of cancer may take even 15 to 20 years to appear after radiation, these expert bodies still were refusing to consider additional cancers even though they realized it might still be too early, 15 years after radiation, to perceive the full effect.

For most of the serious environmental poisons, cancer at 5 to 25 years after the poisoning is precisely the kind of effect we must worry about. Genetic effects, occurring in subsequent generations, can be many times more serious than cancer! The folly of looking for "immediate" effects, and thereby exonerating a poison, must be strongly condemned if disaster is to be prevented.

The Careful Studies Required to Observe Effects

We have pointed out that one person out of 600 dying annually from cancer represents a "major killer" entity—equivalent to the entire cancer problem in the United States. Could officials miss such an effect for an environmental poison through inadequate studies? The answer is "Yes." In fact, one repeatedly encounters supposedly scientific studies that have led to erroneous results and enormous public-health blunders, simply because an inadequate number of exposed persons were studied.

How many people exposed to an environmental poison, radioactivity or other, would be required? At the outset we realize that, whatever the number, the observations won't begin to be meaningful for at least 10 years, because that is when the cancers are beginning to occur in appreciable numbers.

Suppose we ask ourselves about a study involving 1200 people exposed to an environmental poison, a poison that might kill, after a 10 year latency period, one out of 600 people per year. Obviously the first requirement would be a "control" population, *not* exposed to the same poison, to compare with the exposed group of 1200 persons. So we would now be studying 2400 persons, 1200 who had been exposed to the poison, 1200 who had not been exposed.

Let us consider, say, the fifteenth year of the study. Would we easily observe the massive effect that could be occurring? If the spontaneous, or natural, cancer occurrence rate is one per 600 per year, we would *expect* two cases in one year of observation among the control group. We would *expect* four cases in one year of observation for the exposed group (two spontaneously plus two for the poison effect).

BUT there is a random statistical fluctuation in such occurrence rates from year to year, such that the control group might in a particular year show zero, one, two, three, four, or even more cancers occurring. The exposed group, even with four cases expected, may, in a particular year, show anywhere from 0 to 10 cases, with a reasonable likelihood. Thus, what *appears* to be a large study, 2400 persons under careful observation, turns out to be inadequate even to discover an effect so large as to be representative of the entire size of the U.S. cancer plus leukemia problem.

Even if we had 3,000 people in the poison-exposed group and 3,000 in the control group, with the expectancy of 10 and five cancer cases in one year respectively, we would still be uncertain of the meaning of the results, simply because of statistical fluctuations. *At least* 6,000 exposed persons and 6,000 controls would be necessary for a reliable reading of even a *massive* effect—one in 600 per year!

Yet technology promoters seem oblivious to the requirements in terms of how many people must be observed for meaningful answers. They consider a totally inadequate number of cases and conclude "no effect observed." While this may delude the public into accepting poisonous byproducts of the technology, it is not consistent with survival of humans on earth.

We have been discussing here a massive effect that can only be considered a bludgeoning of the human species. And even so, it is apparent that an *inadequate* study can lead, easily, to the ridiculous assertion, "no effect observed."

Atomic energy development must, unfortunately, be regarded as one of the worst examples of irresponsibility of this sort.

The important purpose of demonstrating the rash unsoundness of "expert" pronouncements in the past is to alert the public to such errors so that they will insist on a vastly improved performance in public matters in the future, for radioactivity and other serious pollutants.

In recent testimony before the Congress (Joint Committee on Atomic Energy Hearings) Dr. Paul Tompkins, Executive Director of the Federal Radiation Council, described, with apparent pride, the history of so-called "Radiation Protection Standards." (A more apt description might be the history of "Radiation Disaster Standards.") Dr. Tompkins related that in 1954 the National Committee on Radiation Protection, a leading U.S. group of "experts" had issued the following statement,

> "We have a lower limit of continuous exposure to radiation that is (unavoidably) tolerated by man. There is, on the other hand, a much higher level of exposure that is definitely known to be harmful. Be-

tween these two extremes there is a level of exposure, in the neighborhood of 0.1R per day, that experience to date shows to be safe for the individual concerned."*

Not a shred of scientific *evidence* was produced to support this statement, an astounding statement of supposed reassurance. Now, let us consider, in the light of our medical knowledge of radiation injury 16 short years later, what the cost would have been for public exposure to radiation at such supposedly "safe" levels.

The NCRP was reassuring about 0.1 rad per day. For estimations of cancer risk we customarily estimate the dose for persons of about 30 years of age. At 0.1 rad per day, a person would accumulate 1095 rads by 30 years of age (since 36.5 rads would be accumulated in one year at this rate, it would be 36.5 x 30, or 1095 rads in 30 years). Now, from extensive studies concerning the cancer-producing and leukemia-producing ability of ionizing radiation in humans, it appears that approximately 50 rads of accumulated exposure will *add* as many cancers plus leukemias as occur spontaneously due to "natural" or "spontaneous" causes. And such added cancers and leukemias will occur *each* year for many years once the latency period is over.

By simple arithmetic, 1095 divided by 50 equals 21, so we can expect twenty-one times the natural incidence of cancer plus leukemia. TWENTY-ONE TIMES the natural, spontaneous fatality rate from can-

---

Reference: Dr. Paul Tompkins, quoting directly from 1954 NCRP Statement. In "Environmental Effects of Producing Electric Power." Hearings before the Joint Committee on Atomic Energy," 91st Congress, 1st Session, October-November, 1969. Part I.

cer plus leukemia would have been the result of a dose pronounced by a *body of experts as being without physical effects upon the person exposed.*

No disaster in man's health history could match this one had people truly been exposed to this radiation dose, stated to be *safe* by a standard-setting body, the *National Committee on Radiation Protection.* It is something of a stretch of public credulousness and confidence to call for lasting faith in such "standard-setters."

Earlier, we spoke of the horrors of increasing cancer plus leukemia to double the spontaneous occurrence. The NCRP "safe" dose could have provoked a catastrophe 21 times larger than that!

And this is only the beginning of the incredible fiasco of "standard-setting" for technology. Up to now, we have considered only the cancer plus leukemia part of the hazard. Everyone concerned about radiation hazards to man knows that the genetic consequences in future generations give every expectation of being far more severe than the cancer plus leukemia risk in the current generation of humans.

The United Nations Scientific Committee on the Effects of Atomic Radiation (UNSCEAR) has indicated that it takes about 10 to 100 rads to double the spontaneous rate of genetic mutations. Professor Lederberg, the eminent geneticist, has recently estimated approximately 50 rads to double the mutation rate. His estimate is almost precisely in the center of the range estimated by UNSCEAR, so we may explore the consequences of this estimate. At the average reproductive age of 30 years, a person receiving NCRP's "safe" 0.1 rad per day would have accumulated 1095 rads, we saw above. So, if the genetic mutation rate is doubled by 50 rads, it is increased 21 times by 1095 rads. So the NCRP "safe" dose of 0.1 rads per day would have meant a 2100 percent increase in mutation rate. Contrast this with Professor Lederberg's recent admonition that society would be well advised *not* to add one percent to the mutation rate. Contrast 2100 percent with one percent!!!

It is only by a quirk of fate and timing that society escaped acceptance of the National Committee on Radiation Protection's recommendation, and its results. The nuclear electric industry and other atomic energy programs just weren't ready, technologically, for widespread expansion in 1954.

In the years shortly after 1954, scientists began to wake up a *little* and realize the enormity of the error represented by the pronouncement of the NCRP. It was obvious that a massive reduction, in doses to be allowed for humans, must be made immediately. Biologists in the 1956-58 period, realizing the enormity of their past error, had an opportunity to implement a sound policy with respect to allowable radiation dosage. But they did not do so.

A sound policy of public health protection gave way to the powerful imperative of "convenience" for the promoters of technology. The scientists asked themselves, instead, how low they could push the allowable radiation dose to the public, without interfering with the "orderly development of atomic energy." So they issued suggested standards along the following lines:

5.0 rads per year for workers in atomic energy

0.5 rads per year for individuals in the population-at-large

0.17 rads per year *average* for the population-at-large.

What a come-down these numbers represent from 36 rads per year being "safe" or "without physical effect"!

Incredible as it seems, scientists, in a few short years, had to change a recommendation, downward, between seven and 200 times. And what evidence did these standard-setting scientists provide that the "new" standards of allowable radiation would be safe? None whatever. *Absolutely* none.

Obscurantist gobbledygook has characterized *all* efforts to set so-called "safe" or "allowable" standards for industrial poisons, radioactive or other. In truth, standard-setters know full well there is no evidence for any safe amount of a poison such as radiation or radioactivity.

We are perfectly happy to consider errors of the past as part of the learning process. But the "standard-setters" are not satisfied to learn by errors; they defend their errors of the past and try to justify their unbelievable errors of the present.

For example, the catastrophic statements of NCRP in 1954 are explained this way: "We were not recommending that people be exposed widely to 0.1 rad per day." Thank heaven for this! But, of what earthly use is a pronouncement by a standard-setting body, that 0.1 rad per day is without physical effect upon the exposed person, *other* than as guidance for technologists so they can plan their designs, including safety features?

When the nuclear electricity promoters are asked about hazards due to irradiation at the "allowable" doses of radiation, they go into speeches about the "expert scientists" who set these "allowable" (inferring *safe*) doses after careful deliberation. Indeed, the electric-utility industry buys two-page advertisements in national magazines to present precisely this justification for safety of the "allowable" doses.

When the evidence is presented to the "standard-setters" that large numbers of cancers, leukemias, and genetic disorders would accrue from population exposure at the "allowable" dose, they answer, "We didn't mean for people to ever reach those allowable doses."

Recently, the charade has assumed even more ridiculous proportions as the nuclear electricity salesmen have attempted to defend their obviously indefensible standards for human radiation exposure. An attempt at justification, bizarre in the extreme, is now presented for the 0.17 rads allowable for the population-at-large. It is known that natural sources of radiation *plus* those from medical uses of x-rays add up to approximately 0.17 rads per year. "Aha," say the proponents of nuclear electricity and other nuclear energy programs,

"We shall allow the 'peaceful atom' to give an amount *additional* that will just equal what people are already getting from other sources." But why would anyone think of *doubling* the harm already being produced by the 0.17 rads from natural plus medical radiation? From all that has already been discussed, we know that natural *and* medical radiation produce cancer and genetic harm, in direct proportion to the dose received, down to the lowest doses.

No amount of ionizing radiation is safe!

CHAPTER 5

Promises, Promises

The Congress of the United States, acting in the best of faith during the immediate post-war years, made an historic error in assigning duties and aims to the newly established U.S. Atomic Energy Commission. Atomic energy represented a poorly-understood, new, potent phenomenon, born during World War II. The possibilities and the hazards appeared staggering.

It seemed logical, in 1946, to organize a civilian Commission assigned to explore and exploit the phenomena of atomic energy for the fullest benefit of the citizens. The Atomic Energy Commission was given this as one of its missions. But the staggering potential hazard was also recognized and a second mission, that of proceeding with the fullest consideration of protection of health and safety of the public, was also assigned to the Atomic Energy Commission.

In this dual mission lay the historic error. No group of people could be expected to do both things at the same time—promote a technology zealously and hastily—and at the same time proceed slowly and cautiously for maximum protection of public health. Go fast but go slowly! This was in essence the directive given the AEC at its inception.

As the Commission explored the peaceful possibilities of the atom, one prospect seemed inordinately attractive: utilization of the enormous energy of uranium fission to produce heat, hence steam, and to use the steam to drive electrical generators. The nuclear reactor derives its energy from nuclear fission, rather than from fossil fuel, to produce steam—provided everything goes exactly as planned.

Unfortunately, at several steps along the way, radioactive substances, produced as waste by-products in nuclear reactors, are released into either air or water. The nuclear reactor itself, and the possibility of harm from an accident there, are only the beginning of the story.

Huge quantities of radioactivity are produced in the course of nuclear electricity generation. Electrical power production is measured in kilowatts (1000 watts equal 1 kilowatt) or megawatts (1000 kilowatts equal 1 megawatt). A large power station of any kind produces approximately 1000 megawatts.

For a nuclear power plant operating to produce 1000 megawatts of electrical power, we can estimate how much uranium will be needed. From this we can calculate precisely how much of the various radioactive fission products will be produced, including such infamous ones as radioactive iodine-131, radioactive strontium-90, strontium-89, radioactive cesium-137 and radioactive krypton-85. These radioactive byproducts became familiar to us all during the heated debates over radioactive fallout hazards when bombs were tested in the 1950's.

Some of the radioactive byproducts of nuclear uranium fission have very short half-lives, others very long. This concept of "half-life" seems difficult. It is not. It's mostly just a convenient way to measure the potential for harm and how long it may last. If a radioactive substance has a half-life of one day, we mean that, in the course of one day, half of that substance will decay or disappear. In the next day, one half of what is left will disappear, in the next day one-half of that will disappear, and so on. So a substance with a half-life of one day will be reduced in radioactivity 1000 times in 10 days. Hardly enough left to do much damage, you might say, within the very short time of 10 days.

But if a substance has a half-life of about 30 years (like cesium-137) its radioactivity is reduced 1000 times only after 300 years!

One ugly feature plagues the operation of nuclear

reactors for power generation. As the uranium atoms split, they build up radioactive byproducts which eventually "poison" the reactor itself. Only a small amount of the potentially fissionable fuel can be utilized before it must be removed from the reactor and transported by rail or truck to a fuel-cleaning or fuel-reprocessing plant.

Here the uranium or plutonium is dissolved in acid and purified so that it can be prepared to go back to the nuclear reactor. But astronomical amounts of radioactive byproducts remain, after this process is complete. Usually a nuclear reactor can function for about two years before fuel-reprocessing becomes essential. This means that every two years all of the radioactive material generated by uranium fission must be removed from the nuclear power plant, transported by rail or truck to the fuel reprocessing plant, and there separated from uranium or plutonium which are recovered for future use. The immense quantities of radioactive byproducts must then be transported in some fashion to an ultimate repository.

Plans call for allowing the uranium fuel to remain for a period of months after removal from the reactor so that the short-lived radioactive byproducts decay away. This cuts the radioactivity of the spent fuel rods some, but still massive quantities of the extremely hazardous strontium-90 and cesium-137 have decayed hardly at all in this short cooling-off period of several months.

These radioactive substances, with half-lives of 27

and 33 years respectively, must be kept isolated from the environment for periods like several hundred years if damage to human beings and other living things is to be avoided. It is difficult for the layman to understand or conceive of the enormous quantities of hazardous radioactive byproducts like strontium-90 and cesium-137 that are involved. We will explain.

A 1000 megawatt reactor, operating for two years (the fuel-changing cycle time) produces as much of these long-persisting radioactive poisons as about 2000 atom bombs of the Hiroshima size. This sounds incredible, but is thoroughly documented, as a known fact of physics. Ten such reactors—and the AEC plans for some 500 by the turn of the century—operating for two years have as much radioactivity of long persistence in them as the combined total of such fission-product radioactivities in all the bomb tests of the United States and the Soviet Union combined for the entire period of atmospheric testing up through 1962.

During the bomb tests, that amount of radioactivity spread fallout around the globe, aroused the concern of more than 11,000 biological scientists, and was finally a major factor leading to the 1963 treaty to ban atmospheric tests of such weapons. Yet the AEC is now proposing to build reactors containing inventories many times this total amount of radioactivity on the edge of all our most populous metropolitan centers! Trucks, roaming our crowded highways, will carry radioactive cargoes to reprocessing plants, and eventually to a final burial spot.

Those events which must go absolutely perfectly at every step along the complicated route just described are these:

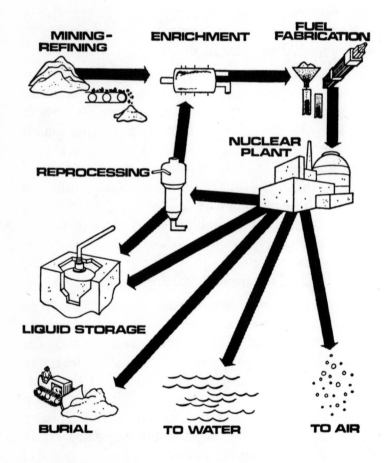

The above diagram shows the course radioactive substances follow from mining through disposal.

112

1. At the reactor itself, bearing enormous quantities of radioactive poisons, no accidents which can distribute such poisons to the atmosphere, land or water can be tolerated.

2. Every two years, the fuel carrying this burden of poison should be transported without mishap by rail and truck to the fuel-cleaning plants. Any significant accidental release in this phase of the operation can render sizeable areas of our nation uninhabitable for many years.

3. At the fuel reprocessing plant absolutely perfect containment must be assured, year in, year out.

4. The waste radioactivities, dangerous for hundreds of years, must be transported to a final resting place. And this waste must be guarded from any escape into the environment for periods longer than the recorded history of any government.

5. At no step (reactor, transport, fuel reprocessing, transport, waste burial) can sabotage of the operation conceivably occur without disastrous consequences for human beings. Yet there will be hundreds of plants and transportation vehicles that must be protected against such sabotage perfectly. Senseless, indiscriminate bombings and arson are hardly an unknown occurrence in the United States today.

We shall return, later, to the issue of a major accident at the reactor itself, and we shall see that no one has the vaguest notion of the risk of an accident there. And we are planning for hundreds of such reactors! Human perfection is required at all these many steps in the entire cycle of events—and required constantly for hundreds of years. No government has ever undertaken such massive responsibility in the history of mankind.

When one considers the fantastic requirements— perfect safety, perfect engineering, perfect reliability, perfect loyalty—for every aspect of such a massive nationwide program to avert disaster, one wonders how the American people can be deceived into accepting such a solution to our power-shortage problems. Obviously, they have no way of knowing any better. They are constantly assured by spokesmen of the AEC and the power companies that nuclear energy is "clean" and "safe."

All that these spokesmen can conceivably mean by the word "clean" is that the radioactive poisons can't be seen or smelled. In many ways it is unfortunate that one can't see or smell radioactivity. If one could, the real hazards of this irreversible environmental poison might be better appreciated by the public.

Instead of considering the multitude of steps that must be carried out perfectly every day, every year, in every reactor they plan to build, in every reprocessing plant, in every truck or railway car carrying radioactive waste and every final burial spot for wastes, AEC officials focus on the very tip of the iceberg by talking only about what radioactivity the reactor itself releases under normal operation.

Precisely how do the AEC spokesmen reassure us

that we won't receive disastrous radiation as a result of the operation of nuclear power plants?

At first, they emphatically denied that 170 milli-rads would produce any significant harm to human beings. They denied and ridiculed the estimates that such exposure of the entire U.S. population could finally produce 32,000 extra cancer and leukemia deaths plus 150,000 to 1,500,000 extra genetic deaths per year. They provided no counter-evidence of their own. They just denied the numbers.

"Scare-laden," the AEC spokesmen proclaimed.

"Alarmist," the AEC spokesman intoned.

"Hyperbolic claims," the AEC spokesmen pronounced.

They offered no counter-evidence. Instead, a steadily increasing number of very prominent biological scientists, not associated with the AEC, announced that the predictions above were by no means exaggerated. Professor Linus Pauling, winner of two Nobel prizes, published his estimate that, if everyone in the country were to be exposed to the allowable amount of radiation, we might expect 96,000 extra cancer-plus-leukemia cases rather than the 32,000 extra cases estimated by us.

Professor Pauling is correct when he states that we estimated 32,000 to be the minimum number of extra cancers and leukemias. Professor Pauling's number, 96,000, does indeed have a high probability of being closer to the true, stupendous cost in human misery and death from exposure to the limits which the Federal Radiation Council has set and which the AEC regards as acceptable.

The eminent Nobel Laureate geneticist, Professor Joshua Lederberg, estimated that the annual cost of the health burden from genetically-induced diseases, at the currently legal Federal Radiation allowable doses, would eventually be 10 billion dollars a year, a number quite consistent with our estimate of 150,000 to 1,500,000 extra genetic deaths per year.

Professor Lederberg added that there were uncertainties in his calculation and that the true financial cost of added medical and health care could range between 1 billion dollars and one hundred billion dollars annually. Even if we were callous enough to disregard the toll of human suffering involved in possibly a million and a half more deaths per year, from degenerative diseases like diabetes and circulatory disorders, the 100 billion dollars estimated as the cost of health care for these unfortunates is roughly comparable to the entire national federal budget annually.

Surely a radiation standard that could lead to this unspeakable burden on society deserves careful examination. Is such a burden necessary for the orderly development of nuclear energy? Indeed, one wonders at the irrationality of such a "standard" for any purpose whatever!

Other scientists, too, have provided their estimates of the cancer-leukemia risk and the genetic risk, including eminent men like Professor E. B. Lewis of

California Institute of Technology, Dr. Karl Z. Morgan, Director of the Health Physics Laboratory of Oak Ridge National Laboratories, and Dr. R. H. Mole of the British Medical Research Council.

Recently Dr. James D. Watson*, another Nobel Laureate in Genetics, stated, "The amount of research now being done on the connection of cancer and radiation is totally inconsistent with proposals for widespread introduction of nuclear power plants into highly populated areas."

Even more alarming than all of these estimates of a high risk of allowable doses of radiation, the great British researcher in the field, Dr. Alice Stewart, came forth with solid evidence that the fetus in utero is especially sensitive—about 50 times as sensitive to cancer or leukemia induction as is the adult.

The precise numerical results for cancers or leukemia predicted for exposure to an amount of radiation proclaimed by Atomic Energy Promoters to be "without effect" differed among the various scientists who provided their estimates. *But all* the estimates pointed to a grave hazard. Is it really very comforting that we estimate 32,000 extra cancer deaths while Professor Pauling estimates 96,000? The real issue is that the hazard is estimated in the many tens of thousands of unnecessary cancer and leukemia deaths each year rather than near zero or at zero.

---

*James D. Watson Testimony before the Hearing Board in the Lloyd Harbor Intervention on the Construction of the Shoreham, Long Island Nuclear Power Station (1970).

Faced with an ever-increasing number of similar estimates of the grave hazard of ionizing radiation, both with respect to cancer plus leukemia and genetic diseases, the Atomic Energy Commission proponents began to realize that their attack on those who estimated the true hazards of radiation was backfiring badly.

So AEC spokesmen began to say, instead, that we —and presumably all these other specialists who have spoken out—just don't understand how the FRC regulations work! Nuclear power plants are not exposing anyone to anything like this amount of radioactivity, they say and therefore, the estimates of the serious hazard of radiation must be wrong.

They appear to have little respect for the intelligence of the American people. What AEC officials are saying, in essence, is that if you are not exposed to allowable amounts of radiation, you won't suffer such devastating effects as cancer, leukemia or genetic diseases. But what are we to make of the AEC's official position that they *will* permit us all to be exposed to this limit of 170 millirads per year and that they believe no harm can come to us if they do?

AEC officials point out that nuclear electric power generation hasn't yet delivered anywhere near the 170 millirads, as an average, to the population of the United States. So, fortunately, the American people have not, as yet, been exposed to highly dangerous levels of radioactivity. But the AEC fails to point out that nuclear electric power stations haven't generated enough electricity, so far, to be worth discussing. A handful of small nuclear power plants is in operation —the largest approximately one-half the power level of those being planned.

But now the AEC is planning, ultimately, 450-650 large nuclear power plants, plus the necessary reprocessing plants, plus the necessary transportation and burial facilities—roughly a thousandfold increase in nuclear power generation! And, incredibly, they ask us to believe that with this thousand-fold increase in potential radiation hazard—we are not likely to experience any more exposure to radiation than during these early days when the nuclear industry is in its infancy.

This is like saying that a thousand eight-cylinder cars, packed into a mile of highway, are not likely to produce any more air pollution than one model-T Ford!

Spokesmen for the nuclear power industry assume that all nuclear power plants would operate perfectly according to design specifications which they expected to be infallible. They ignored prior experience, which shows this to be a pipe dream.

Second, they conceive only a minute possibility of minor or massive accidental releases of radioactivity at the nuclear power station or in the transport of the radioactivity-laden fuel rods, or at the reprocessing plant, or in the preparation of the mammoth quantities of radioactivities for ultimate burial, or in the transport of such enormous quantities of radioactive debris to ultimate burial sites.

119

Third, they assumed that sabotage at any step along this chain to be unthinkable. Largely because the thought was so chilling, the AEC officials hoped the thought would go away.

Fourth, for some reason they chose to ignore a major pathway for delivering serious doses of radioactivity to man—the processes by which plants and animals in the food chain of man can concentrate radioactive substances in a massive manner.

They say the radioactivity release from reactors now is, and in future will be, only 1 per cent of the official guidelines. We shall discuss this optimistic statement in more detail in the next chapter. The guidelines they refer to here are the Maximum Permissible Concentrations in Air and the Maximum Permissible Concentrations in Water.

In the Code of Federal Regulations, Title 10, pages 134 to 144, is a Table listing the maximum permissible concentrations, of various radioactive substances in air and water, which are permitted to be released to an unrestricted area—that is, any part of the community outside the confines of the nuclear plant itself. (Title 10, CFR, part 20.)

These levels are set so that a whole-body dosage of 0.5 rads per year would result from breathing such air for one year, or drinking about two quarts of such contaminated water per day. But what do such levels really mean in terms of what could occur, and probably will occur if such levels are allowed in an unrestricted area where people live?

Cesium-137 in the air near the power plant will deposit on nearby pastures. This will be grazed by cows and the cesium-137 in their milk will eventually be consumed by children. If we allow the permitted level of cesium-137 concentration in the air for just one day, a child consuming one liter of milk every day will get a whole-body dose of seven rads *as a consequence of just one day's exposure.*

If the Maximum Permissible Concentration of cesium-137 in the air is maintained for one year, the dose will be 2,555 rad which is 5,110 times higher than the 0.5 rad guideline *and 15,000 times more than the 0.17 rad radiation protection guideline of the Federal Radiation Council*—not from the air the child is breathing—but from the milk he is drinking!

Let's look at the concentration in water. The MPC is based upon the calculation that a 150-lb. standard man consuming 2200 grams of water at the MPC per day would receive a dose of 0.5 rad. To begin with, a 75-lb. child drinking this much water would get a dosage twice as high. He would be exceeding the guidline dosage and so would a 100 lb. pregnant woman. Man, woman and child have also been known to eat fish. The concentration of Cs-137 in fish flesh, caught in a river, would be 1000 times higher than the concentration in the water. Thus a man eating 1-lb. of fish a week, grown in water at the MPC, would receive a dosage of 15 rad/yr or 30 times the 0.5 rad guideline and 90 times the 0.17 rad guideline. If he were a 75-lb. child, the dosage would be 60 times the 0.5 rad guideline and 180 times the 0.17 rad guideline. In other words, most people would exceed the guidelines if they ate only one pound of fish a year.

The milk and fish represent biological concentration mechanisms. They, by themselves, serve to demonstrate quite conclusively that using air and water MPC values without considering food chains is meaningless. But as another example, let's look at a physical process. If the Cs-137 MPC in air were maintained for one year, the radiation level would be 23 rad per year.

Thus, when the AEC officials state that releases will be only 1 percent of the guideline, we shouldn't be lulled into complacency. The above example for Cs-137 in milk indicates that, for the 0.17 rad guideline, the releases should be 0.007 percent of the MPC_a, not 1 percent. If a more reasonable primary standard of 0.017 rad were applied, the allowable release would be only 0.0007 percent of the MPC_a, more than a hundred thousand fold lower than the current MPC for air.

The AEC officials only look downwind from the plant for people breathing in the *air* containing the radioactive cesium and they neglect totally that the contaminated milk described above can be shipped hundreds of miles away and deliver large doses to the residents of a major city nowhere near the reactor.

Thus, by neglecting *all* the important routes by which radioactivity from nuclear power plants, transportation of radioactivity, and from fuel processing plants and ultimate waste storage gets to people, the AEC officials conclude that, in the foreseeable future, no one will be exposed to anywhere near the allowable radiation dosage.

We can easily test whether AEC spokesmen really believe what they say, as they vie with one another to see who can make the rosiest predictions.

Commissioner Thompson, in a recent speech,* stated that:

> "As I have already indicated, it is likely that even by the Gofman hypothesis (that 170 millirads to the entire population will lead to 32,000 extra cancer and leukemia deaths annually) less than one person per year would be in jeopardy due to the presence of reactors compared with a sum total of 300,000 cancers per year from other causes.

> Dr. Thompson became even bolder in his following statement.

> "Instead of having 32,000 cases per year, we probably have statistically less than one extra case of cancer or leukemia as a result of the presence of those nuclear reactors now in operation, construction or definitely planned."

What, in effect, has Commissioner Thompson commited himself to and can he make good the commitment he so casually makes? He asks us to assume that we are correct in our prediction of 32,000 extra cancer deaths if the average exposure of everyone in the country is 170 millirads per year. He also assures us that future reactor programs will not result in more than one extra cancer death per year. This means he

---

*"Power Technology and The Future", AEC Commissioner Theos Thompson. Delivered at "Briefing Conference for State and Local Government Officials on Nuclear Development", Columbia, South Carolina, May 21, 1970.

is willing to guarantee that the average dose of radiation to the American people will be 1/32,000 of the currently allowable dose, even after another 500 or so nuclear reactors are spread all over the landscape! He guarantees a dose of about 0.005 millirads.

The AEC commissioners know perfectly well that it is meaningless to discuss only radiation from the nuclear reactor itself. They are here assuring us, in the words of Commissioner Thompson, that the combined radiation dosage, from the reactor, from transporting spent fuel rods, from reprocessing fuel, from radioactive waste preparation and storage of waste for all centuries to come — including any and all accidental releases—will be less than 0.005 millirads per year for the American people.

We would be delighted if the AEC and the electric power industry could make good on this promise, which is made, remember, by the men assigned by the federal government to protect us all from radiation hazards. If the AEC could, indeed, assure the American people that the development of nuclear power plants, in the numbers which they have promised us, can be accomplished without exposing all of us to more than 0.005 millirads of radiation a year, critics of the nuclear power program would certainly withdraw their criticism and expressions of concern and alarm.

But when we challenge this statement, by asking that the official radiation exposure level be reduced to 17 millirads or less, AEC officials call us alarmists and insist that nothing of the kind is necessary. Does it not seem strange that they claim they can develop a widespread nuclear power industry without any possibility of exposing us to more than 0.005 millirads a year, but, when we ask for a reduction in allowable standards to a value *3400* times this high, they say they cannot allow it.

They claim that a little leeway is needed for unexpected incidents. Surely "unexpected incidents" do not require 3400 times as much possible exposure— which they characterize as "a little leeway." We might understand two times or even ten times the guaranteed level, but 3400 times strains our credulity.

Commissioner Thompson said that present reactors do not account for even one cancer death per year, which implies that the present exposure of the entire population is 0.005 millirads or less. Yet the director of the Federal Radiation Council, Dr. Paul Tompkins, stated that it would cost billions of dollars to rebuild reactors now in operation to comply with an allowable dosage of 17 millirads per year*. If, indeed, the current exposure is as low as the AEC claims it is (3400 times lower than 17 millirads) we shouldn't need any revision of reactor installations at all and the cost would be zero.

It's obvious that Commissioner Thompson's estimates, which we presume are the official estimates of the AEC, differ from those of Dr. Paul Tompkins by factors of many thousands. How does this happen?

---

*"U.S. Responding to Radiation Warning" by Roger Rapoport, San Francisco Chronicle, December 18, 1969.

And how can the layman, with no expertise in these matters, have any confidence in what such public statements may mean, when the experts differ so radically?

Dr. Victor Bond of the AEC's Brookhaven Laboratory, makes even rosier predictions than did Commissioner Thompson. Dr. Bond testified at recent hearings before the Public Service Board of the State of Vermont. His written testimony was from a document entitled, "The Public and Radiation from Nuclear Power Plants."**

Dr. Bond, too, sees only the tip of the iceberg — the nuclear reactor itself, operating perfectly, with no radiation from mechanical failures, no accidents, no carelessness, no mistaken judgment on the part of employees. He does not once mention the chance for radiation exposure in all the other aspects of nuclear power which we have described: transporting fuel rods, processing fuel rods, transporting wastes, storing wastes.

He estimates that nuclear power plants at present expose the American people to an average of 0.0001 millirads per year. He calculates that, for a forty-fold increase in the nuclear power industry in the future, this exposure might go to 0.004 millirads per year.

Therefore, he reasons that estimates of cancer-leukemia risk or genetic hazard based on the currently allowable 170 millirads are some 42,000 times too high. Such an assumption implies that nuclear power

---

**The Public and Radiation From Nuclear Power Plants, Victor P. Bond, Testimony delivered by the Public Service Board of the State of Vermont. Hearings on the Vermont Yankee Nuclear Power Plant, September, 1970.

plants for electricity generation can proceed to expand fully as planned, up to the year 2000, with exposure limits 42,000 times lower than those currently allowed!

Such an assumption implies that any radiation exposure from fuel transportation, accidents, sabotage, fuel processing, waste radioactivity processing, waste radioactivity shipping and perpetual guardianship of immense amounts of radioactive waste will be totally negligible in comparison with the 0.004 millirads predicted as the dose expected for the American people from the nuclear power plants alone — the tip of the iceberg.

At the hearings, Dr. Bond was asked by Attorney Bloustein why he opposed lowering the allowable amount of radiation when he claims there is a 42,000-fold difference between what he, Dr. Bond, believes is required and what is now permitted by Federal Statutes. Dr. Bond was unable to answer.

So, in hearings throughout the country, in speeches before many varied groups, and in testimony before Congressional committees, we hear AEC spokesmen and promoters of the nuclear power industry trying to outdo one another in their predictions of how low the radiation exposure of the American people will be for all aspects of nuclear power generation — 30,000 to 40,000 times lower than the levels currently allowed.

Yet when highly competent biologists in this field ask that currently allowable doses of radiation be reduced only 10-fold or 100-fold, proponents of nuclear electricity refuse to consider this change, even though it would give them a 300 to 3000-fold margin of safety above what they promise will be the average radiation exposure.

One could be very generous in providing leeway for "unexpected incidents." We could allow the nuclear-power industry to develop with a radiation dose allowance (including all hazards of the industry) of 0.1 millirad for the population. This provides a 25-fold margin of safety over what Dr. Bond says is required. This is very generous leeway. If the nuclear power industry were to accept this level, all the arguments would end and the nuclear industry could proceed unhampered.

Of course, these arrangements would have to be entered into the Code of Federal Regulations. Present promises and good intentions are not. One cannot place promises and good intentions in a Code of Regulations. Yet, in a matter involving irreversible pollution of the human race and the environment, the proper and only place for all agreements in regard to this matter is the Code of Federal Regulations.

One must reluctantly conclude that there is a great deal of confusion and lack of responsibility at very high levels in the entire program. The AEC was presented with evidence that all the standards they had proclaimed as safe were truly unsafe.

The atomic energy proponents vigorously denied that harm in the form of cancer, leukemia and genetic diseases was even possible.

Representative Chet Holifield, chairman of the

Joint Congressional Committee on Atomic Energy, told us he had been assured that 100 times as much radiation as the level which is officially allowable would be necessary before the safe level is passed. Chairman Holifield's statement rests on totally discredited evidence. The AEC statement that no effects are observed at their presumed safe levels of radiation means only that no one has ever made any adequate observations.

When it became clear that unproven claims of the AEC and the Federal Radiation Council were being exposed, the second line of defense was used, the so-called "safe" standards must be safe because very sincere men had set the standards. These same men now refuse to look at the massive new evidence which proves their standards to be anything but safe.

When this strategy of denying the evidence was ineffective, promoters of the nuclear energy program took up a new tactic. "We will never allow anyone to be exposed to the allowable dose," they said. But when they were asked for minimal evidence of sincerity in the form of a written regulation in a government code, to guarantee the public that the nuclear power industry will not expose us to the maximum level, they refused absolutely.

It seems that the public can draw only one conclusion — that neither the AEC nor the nuclear power industry believes they can operate at the low doses they promise. So they hope against hope that the public will not require them to make good their claims and will, instead, accept their "promises, promises."

CHAPTER 6

How Safe Are Nuclear Reactors?

Routine Operations

Discussion of the safety of nuclear reactors has two components: (1) The normal day-to-day operations in which the reactors are permitted to release radioactivity into the environment and (2) the accident situation wherein the reactor may release large quantities of radioactivity into the environment.

In normal day-to-day operations, nuclear power plants are permitted by law to release radioactivity in the form of radioactive atoms to the environment in gaseous and liquid discharges. There are essentially two regulations concerned with these releases. The regulation which represents the primary standard is the dosage that could be delivered to an individual or to the population-at-large. We have discussed this primary standard earlier in this book and have indicated that the standard is much too high. The maximum permissible concentrations of the various radionuclides in air (MPC$_a$) and water (MPC$_w$) that are permitted to be released outside of the restricted area of a nuclear reactor are called secondary standards.

The primary standard should be derivable from the secondary standards. But, as we indicated earlier, the secondary standards, the maximum permissible concentrations that are listed in Title 10 of the Code of Federal Regulations, do not permit this. The MPCs that are tabulated in Title 10 of the Code of Federal Regulations apply only to the situation where individuals are breathing the contaminated air or drinking the contaminated water. They do not take into account the fact that the contaminated air and the contaminated water will result in the contamination of the foods consumed by man. This is an extremely important factor in terms of the dosage that would be received by man from reactor releases.

The proponents of the nuclear power industry state that the exposure from the nuclear power plants would be considerably lower than those of the guidelines. They indicate that individuals in the near vicinity of nuclear reactors would be exposed to no more than 5-10 mr/yr value and that as individuals lived further and further away from the reactor, their exposure would drop off very rapidly from this 5-10 mr/yr value. Moreover, they indicate that the design objectives and the operation of existing power plants are such that the actual releases of radioactivity from the power plants are no more than 1 percent of the releases allowed by the AEC's MPC values.

The State of Minnesota has proposed emission standards that are 50 fold lower than those of the AEC. At a symposium titled "Nuclear Power and the Environment", held at the University of Minnesota in October, 1969, members of the audience repeatedly asked the question, "If reactors are only going to release the small amount of radioactivity that you indicate, then why are you so reluctant to make guidelines more restrictive and adopt the Minnesota regulations?" Congressman Craig Hosmer, a member of the Joint Congressional Committee on Atomic Energy, stated that if the standards were lowered, he doubted if the reactors could operate safely. Commissioner Theos Thompson made essentially the same statement in testimony before the Joint Committee on Atomic Energy.

In recent hearings before the JCAE, the following exchange took place between AEC Commissioners James Ramey and Thompson and Congressmen Chet Holifield and Hosmer. We will reprint their exact words, with only a bit of interpretation.

> Chairman HOLIFIELD . . . One other point I wanted to bring out was that the newer plants and the plants that are now being put on the line commercially and which do not have experiments involved in their

continuous operation show consistently a concentration limit of less than 1 percent.

Is that not right, or am I wrong in reading the chart?

(Holifield means that the radioactivity escaping from the plant is less than 1 percent of official limits.)

Dr. THOMPSON. That is correct. You are reading the chart correctly.

I would like to make the statement, though, that there may be times, when, even in spite of careful inspection—which is always done—and the checkout to assure that the surface of these fuel elements is free from uranium, the effluent levels will rise above this one percent but still be well within the current part 20 limits.

It is therefore important that we have what I will call an operating cushion.

You talked with Dr. Totter the other morning and you asked him whether there was a reasonable cushion of safety between the effects of radiation and the present standards which are set up.

What I am talking about is another cushion between the part 20 standards and the normal operating level. That cushion is important for the reliability of these plants as a part of an electrical utility system. Assume a utility builds a nuclear reactor, and then, say, they go up from one percent to three percent of the part 20 limit. Then, if we have set a lower limit at one percent, this reactor would have to be shut down. But it would not really be shut down because of a safety reason but simply because somebody had arbitrarily established a very low limit.

(Dr. Thompson means that, if the limits of radiation at the reactor were set at 1 percent of the part 20 limit,

134

but the operators found they could not keep the limits this low, then the reactor would have to shut down.)

This would materially reduce the reliability of this plant as a power source.

I think the AEC has an obligation, as a responsible group, to be sure that the reliability of these plants is not reduced by making these standards too low just arbitrarily.

(By reliability he means here its ability to continue to provide power. He is not talking about reliability in terms of safety.)

Chairman HOLIFIELD. Of course, if safety standards are too restrictive, the various attempts to comply with a too rigid standard would increase the probability of trouble from a technological standpoint; is that not true?

Dr. THOMPSON. If the radioactive effluent standards are too rigid, in my own mind at least, there are some very grave worries that I have concerning whether this may not reduce the ultimate safety of the reactor plant itself. If one begins to push too hard on holding down effluents, one may as a result affect reactor safety adversely. For instance, say we hold up all the tritium in the plant. This tritium makes very high levels of tritium in the air inside the containment. Then the tendency will be not to inspect the plant so often.

Another example. In the boiling water reactor, there are those who would cut down the effluent that is released through the stack too strenuously and too early before technical feasibility for doing this has been demonstrated. It may well be that, as one moves to a very long holdup of gases in the boiling water reactor effluent system—and a lot of the gases which come out from this plant are really hydrogen and oxygen which are disassociated in the core of the re-

135

actor—there is a possibility that unless one is very careful you will induce an explosive hazard where no hazard was there before.

Therefore there is a very close interaction between effluent discharge levels and safety of the reactor.

I am somewhat concerned that we will move from a more safe reactor to a less safe reactor if we push the effluents down more than we should on a reasonable basis. I believe we are on a reasonable basis right now.

(In other words, the more you restrict the levels of radioactivity loosed on the world outside the plant, the more you risk a possibly catastrophic explosion at the nuclear reactor.)

Representative HOSMER. You brought up this matter of the cushion. You used as an example if you go from one to three percent.

Dr. THOMPSON. I picked the number one to three.

Representative HOSMER. Let us call that a size three cushion. But you go to 100 percent under the same limitations. Would you then be using a size 100 cushion?

Dr. THOMPSON. That is right.

Representative HOSMER. So, let us get into the reasonableness of the size of the cushion. We know that the limitations are established on the basis that you can go up to 100 and still do no damage to individuals and the public but some people seem to think that there is not enough known so that that might not be an absolute guarantee.

So, why don't we think in terms of reducing the legal size of the cushion to what would be reasonable?

If you say you want to go up to three percent, maybe size three cushion or maybe size 10 cushion, to give you some extra latitude, some elasticity, you

know, to assure the public again and again and again
that their health and safety is being cared for?

(We think Rep. Hosmer's statement here indicates,
perhaps better than anything else could, the total confusion that exists in regard to the possible hazards we
face. An important question is why do the Joint Committee and the AEC assure us that we are in no danger,
even though they themselves confess to a great deal of
confusion and uncertainty?)

Mr. RAMEY. Mr. Chairman, may I comment on
that?

Chairman HOLIFIELD. Yes; proceed.

Mr. RAMEY. I think we do have the standard
for guidance here and it is the standard that is under
the FRC of holding the levels as low as practicable.

We have looked at this rather carefully. We are
still looking at it as Dr. Thompson has indicated, but
there are these factors that we have to take into
account in balancing this, these trade-offs between
reactor safety and the safety from the effluents. It
might be possible to give some guidance as to what is
practicable, how this could be handled. But it is not
likely to be something that sets some limit in terms of
radioactivity. It is more likely to be guidance in terms
of design and in terms of operating procedure on how
the utilities now are holding these levels down in these
ranges. Because every once in a while you may have
to go up above any particular limit and be near your
100 percent factor.

Representative HOSMER. Mr. Ramey, with the
older reactors that Dr. Thompson has just discussed,
the Humboldt Bay reactor, for instance, the technology has now proceeded to where the practical
limits observed in the normal course of operation are
by a factor of 100 below the legal limits prescribed
in the licensing process.

Since the technology has developed and since the practical limits are being observed, all I am trying to seek is some accommodation between the present legal limit and the practical limit at which the elasticity, the cushion, would be adequate but at the same time the legal limits reduced.

Mr. RAMEY. As I say, Mr. Hosmer, I think if we look on part 20, that number is initially set, as has been brought out, as a very conservative number in the first place with a great number of factors of conservatism in it.

Representative HOSMER. Part 20 has already been described by Dr. Thompson as a "dynamic and living thing."

Mr. RAMEY. That is right. I think the way we are looking at it is in terms of within it, and in accordance with the FRC guidelines, of how one might provide guidance on what is practicable below the part 20 numbers.

Now, operating experience has shown that these are at a fraction of the part 20 numbers on these types of reactors.

There are transient situations which may exceed this experience. For example, in Minnesota, in this Minnesota permit, what they have taken as an average and made that the limit. Anybody knows when you set a limit based on an average that sometimes you are going to go over that average and at other times you are going to go under it.

So, if you set it at the average and as an absolute limit, you are going to be violating it.

Representative HOSMER. I am not talking about an average. I am talking about an average plus a reasonable cushion and asking if size 99 to 100 is a reasonable cushion or wouldn't size 25, for example, be a reasonable cushion.

Dr. THOMPSON. Mr. Hosmer, we don't have at

the moment any way to set a reasonable cushion. There is not that sort of experience. So we should not move and make that cushion smaller until such experience exists.*

Basically what the AEC Commissioners are saying is that they don't want to change the standards until they know how much radioactivity will be released. If the reactors are going to release only 1/100th of the present allowable release rates, then why should the AEC be so reluctant to lower the standards by at least a factor of 10? *The only conclusion that a reasonable person can come to is that the AEC does not firmly believe that the reactors will be able to operate at these lower release rates.*

This 1 percent release rate is a design objective. Dr. Thompson recognized this in his testimony and also recognizes that an operating plant may exceed these objectives by a wide margin. A little later on in his testimony, Dr. Thompson stated:

> Frankly, at this stage in the development of atomic energy I think it would be premature to set this, say, three percent cushion or 10 percent cushion, in an arbitrary manner. I think we ought to take a look at the large plants that are coming on line and see how they are going to do. I think they will be at the same levels as present plants but we also need a fair amount of cushion . . .**

Mr. Wilfrid Johnson, another AEC Commissioner,

---

*In *Environmental Effects of Producing Electrical Power.* Hearings before the Joint Committee on Atomic Energy, 91st Congress, 1st Session, held Oct. 28-31, Nov. 4-7, 1969. Washington, D.C., U.S. Government Printing Office, 1969, Part 1, pp. 203-205.
**Ibid.*, p. 206.

supported Dr. Thompson's position a little later in the testimony:

> Mr. JOHNSON. I wanted to add, with regard to the same point that Dr. Thompson brought up that we do need the flexibility in the levels, in part, because they have to apply broadly over various kinds of plants, such as chemical processing plants, as well as reactors. They are also related to occupational exposures.
>
> There is no way to completely divorce the matter of effluents of a plant from the occupational exposure that the employees get. They are related matters.
>
> On top of that, we must consider new plants that come along. They will have different kinds of releases and the limits have to apply to them, too.
>
> If we were too rigid, we would have nothing but boiling water and pressurized water reactors from now on. If we get to liquid metal cooled fast breeders, the effluent problem will be different. Hopefully, they will be better, but we know they will be different. We need flexibility for these reasons.*

(Here Commissioner Johnson admits that nobody knows what the effluent problem will be in the fast breeder reactors which the AEC assures us are the only final solution to our power problems. They have announced plans for such a reactor at Meshoppen, Pennsylvania. Presumably they must wait until this reactor is in operation before they will know how much radiation will be released in its operation!)

To a considerable extent, the amount of radioactivity released to the environment by an operating nuclear power reactor depends upon the integrity of the

---

Ibid., p. 209.

fuel rods in the reactor. The large reactors that are planned, and are being constructed, in this country today have thousands of these fuel rods inserted into the core of the reactor. These fuel rods can develop small pin holes. The radioactivity generated within the fuel rods then leaks through these pin holes and into the water which is moderating the reactor. (See diagram of reactor core on page 21.)

In a boiling water reactor the gaseous products will be released through the stack. The reactor is not able, completely, to contain this water which bathes and moderates the fuel elements and collects the radio-activity which leaks from the rods. Therefore, radio-actively-contaminated water accumulates within the reactor housing. This radioactive waste water is then released into the cooling water and returned to the river or to the ocean.

Consequently, the degree below the maximum per-missible concentrations that a given reactor will be able to operate depends upon the integrity of its fuel rods, as well as the integrity of all the valves, nozzles and pipes in the plumbing and cycling system of the reactor. The reactors presently under construction are planned to operate for some 20 years. Plans are to change the fuel rods only once every two or three years. These reactors are considerably larger than the reactors upon which we have any experience to date.

The combination of these things indicates that we do not really know how these reactors will operate as they begin to age and as their fuel rods begin to age.

It may well be that the natural aging process of the reactor, variations in quality control, and operator errors will cause it to creep up to the maximum permissible concentrations that are presently allowed by the AEC. They might even exceed those levels!

Since nuclear power reactors are being proposed at a rate which indicates they will be supplying a very substantial fraction of our future electrical power needs, we will be presented with a *fait accompli* in the future. Even if these reactors do not operate at their design specifications, it will be difficult to shut them down because we will need the power. If we shut them down, sizeable sections of the country will experience periods of brown-out. We might, therefore, be forced to live with whatever radioactive emissions the reactors require. Once we have made a very sizeable commitment to nuclear-generated power, we must face the fact that we will be stuck with the commitment.

The discussion above indicates that the present generation of reactors is no more than an experiment. The public is told that the guidelines are safe. But they are not safe! The public is told that the radioactive emissions *will be* only 1 percent of the guidelines. This is a design *objective*. An objective that the AEC Commissioners are not in the least certain will be met. The AEC is adhering to its guidelines in order not to inhibit the development of the nuclear power industry by engineering, operational, or quality control failures. The public is simply required to take the risk inherent in this "Cushion."

Accidents In Present-Day Reactors

In addition to the uncertainties in the day-to-day release, uncertainties exist about chances for a major accident. Dr. Walter H. Jordan, Assistant Director of the Oak Ridge National Laboratory and a member of one of the AEC's reactor safety boards, stated in a recent article in *Physics Today:*

> The important question still remains: Have we succeeded in reducing the risk to a tolerable level, that is, something less than one chance in 10,000, that a reactor will have a serious accident in a year?
>
> Have we succeeded in reducing the hazard to such a low level? There is no way to prove it. We have accumulated so far some 100 reactor years of accident-free operation of commercial nuclear electric power stations in the U.S. This is a long way from 10,000 so it does not tell us much.
>
> The only way we will know what the odds really are is by continuing to accumulate experience in operating reactors. There is some risk but it is certainly worth it.*

How safe are nuclear reactors? Let us quote from consulting engineer, Adolph Ackerman:

> As an independent consulting engineer I have been active for many years in alerting the engineering profession to its overriding responsibilities in design and construction of safe atomic power plants. The simple fact is that none of the atomic power plants currently in operation or under construction have been designed with the traditional concepts of engineering responsibility and ethical commitment for maximum public safety.**

---

*Walter H. Jordan, "Nuclear Energy: Benefits versus Risks," *Physics Today* (May), 32-38, 1970.
**Personal communication.

Mr. Ackerman spelled out his reasons for this statement quite clearly in a recent article. Professor Robert L. Whitelaw, of the Virginia Polytechnic Institute and formerly Project Engineer for the design and construction of the power plant for the nuclear ship, N. S. Savannah, commented on this paper by Ackerman in the *IEEE Transactions on Aerospace and Electronic Systems* (vol. AES-5, no. 3, May 1969):

> I wish to endorse fully the principal argument advanced by A. J. Ackerman in his paper and, perhaps, strengthen the impact of his paper with this brief discussion.
>
> His principal argument has been confirmed by my own experience of the past fifteen years on nuclear projects and problems of various kinds. This experience included preparing proposals and nuclear hazards evaluations in a variety of nuclear power plants, both commercial and military.
>
> It has been my observation that, despite the enormous amount of meticulous detail which the ACRS regularly requires on every projected power plant to satisfy itself that there is no "credible accident" that can threaten the public (or even the operators)—and despite the volumes of paper and hours of presentations consumed on this topic, and no doubt well-intentioned—there is still by common consent an unwritten agreement to treat as "incredible" the most fearful of all nuclear accidents that can occur in any plant with a highly pressurized primary system. Such an accident is, of course, the explosive rupture of the primary vessel itself, which is ruled out of the list of credible accidents for the simple reason that there is no adequate answer short of putting the plant underground or inside a mountain, as Ackerman has pointed out.

144

Dr. Edward Teller, often called the father of the hydrogen bomb and one of the most outstanding supporters of the AEC, has stated:

> A single major mishap in a nuclear reactor could cause extreme damage, not because of the explosive force, but because of the radioactive contamination. . . . So far, we have been extremely lucky . . . But with the spread of industrialization, with the greater number of simians monkeying around with things they do not completely understand, sooner or later a fool will prove greater than the proof even in a foolproof system.*

On September 10, 1970, in Livermore, California, Dr. Teller told the Livermore Chapter of the Society of Professional Engineers that reactors were safe, but they should be put underground.

How safe are nuclear reactors? Let us quote from a letter of the AEC's Advisory Committee on Reactor Safeguards concerning a reactor planned for Midland, Michigan.

> . . . The number of permanent residents within five miles of the plant site was estimated to be 41,000 in 1968, mainly in the city of Midland and its environs.
>
> The applicant has established criteria for, and has begun the formulation of a comprehensive emergency evacuation plan . . .

In considering the safety of nuclear reactors, it is important to recognize that each nuclear reactor in this country is an experiment. Each reactor is different from all other reactors and whether or not it will

---

*As seen in Eugene Register-Guard (Oregon), October 7, 1969.

operate and/or operate safely depends upon the outcome of the experiment.

One of the reasons for this is that the AEC has not funded safety research at an appropriate level. This was recently pointed out by Mr. Joseph M. Hendrie, Chairman of the Advisory Committee on Reactor Safeguards, in a letter to Dr. Glenn T. Seaborg, Chairman of the Atomic Energy Commission, dated November 12, 1969:

> DEAR DR. SEABORG: The Advisory Committee on Reactor Safeguards (ACRS) wishes to reemphasize some previous recommendations concerning the need for safety research in several important areas in which the effort has not been sufficient. The Committee has been recently informed that overall reactor safety funding for FY 1970 and 1971 will be considerably below the AEC estimates of need for the water reactor safety research program, as well as for safety research on seismic effects, on sodium-cooled fast reactors, on high-temperature graphite-moderated, gas-cooled reactors, and on environmental effects. As a consequence, many safety research activities have not been initiated, have been slowed, or have been terminated. The Committee reiterates its belief in the urgent need for additional research and development in these areas, and refers in the paragraph below to earlier statements of the Committee on these subjects.

> ### Water Reactors

> In its letter to Mr. Hollingsworth of March 20, 1969, the ACRS stated its belief that ". . . more effort should be devoted to gaining an understanding of modes and mechanisms of fuel failure, possible propagation of fuel failure, and generation of locally high pressures if hot fuel and coolant are mixed, and that effort should commence on gaining an understanding

of the various mechanisms of potential importance in describing the course of events following partial or large scale core melting, either at power or in the unlikely event of a loss-of-coolant accident." The Committee has strongly recommended safety research of this kind several times during the last three years; the Regulatory Staff has also strongly supported such work. However, only small or modest efforts have been initiated thus far.

In its comments of March 20, 1969, the Committee also recommended that ". . . considerable attention be given now to the potential safety questions related to large water reactors likely to be proposed for construction during the next decade. Large cores, higher power densities, and new materials of fabrication are some of the departures from present practice likely to introduce new safety research needs or major changes in emphasis in existing needs.

The Committee further recommended that consideration be given to ". . . research aimed specifically at improving the potential for siting of large water reactors in more populated areas than currently being utilized; for example, studies should be undertaken to develop reactor design concepts providing additional inherent safety or, possibly, new safety features to deal with very low probability accidents involving primary system rupture followed by a functional failure of the emergency core cooling system."

It appears that, because of funding limitations and for other reasons, the recommendations of the ACRS will not be implemented at this time.

Liquid-Metal-Cooled Fast Breeder Reactors (LMFBR)

The ACRS, in its report on safety research of November 19, 1963, stated that "Recent renewed emphasis on the long range role of large fast breeder reactors points up the need for a well developed, long term, comprehensive research program on the safety

147

of such reactors. A strong research program started now should develop information very useful to the first generation of very large fast reactors." The Regulatory Staff and the ACRS have recently undertaken a preliminary review of a proposed site to be used for construction of a 500 MWe LMFBR. Construction permit reviews, of one or more LMFBRs, are anticipated in the next few years.

While an extensive LMFBR safety program plan has been formulated, and a growing program in LMFBR safety has been started, many safety-related design decisions will have to be made by applicants and the regulatory groups without the benefit of needed safety research, in part, because of a lag in the implementation of studies of high priority matters.

. . . In summary, the Committee again emphasizes the importance of safety research to the protection of the health and safety of the public and urges that adequate funding be provided to permit timely pursuit of work in all high priority areas.*

In a letter of November 12, 1969, to Mr. Robert E. Hollingsworth, General Manager of the U. S. Atomic Energy Commission, Mr. Hendrie stated:

. . . The water-cooled reactor safety research program in PBF (power burst facility) should concurrently investigate, with high priority, the mechanisms and phenomena associated with the initiation, growth, and propagation of fuel pin failure, including the circumstances under which melting of fuel could progress beyond one fuel element. Such a situation could develop in a large power reactor because of a local reduction in heat removal rate (as by-flow blockage), a locally abnormal power density (as by incorrect

---

*In *AEC Authorizing Legislation. Fiscal Year 1971.* Hearings before the Joint Committee on Atomic Energy, 91st Congress, 2nd Session, held March 11, 1970. Washington, D.C., U.S. Government Printing Office, 1970, Part 3, pp. 1619-1620.

enrichment of fuel), or a more widespread perturbation in power or flow. These experiments are required in order to ascertain the probability of a local incident progressing into a serious accident and, if possible, the course and consequence of such a sequence of events.*

These complaints, by the AEC's Advisory Committee on Reactor Safeguards, suggest that the present reactors and those under construction are far more experimental than we might have imagined.

It is significant to note, particularly in relationship to the ACRS concern over loss of coolant, which it considers as an unlikely event, and Dr. Teller's statement about "simians monkeying around," that Mr. E. P. Epler discusses an emergency cooling system failure in the Oak Ridge Research Reactor in the July-August, 1970, issue of *Nuclear Safety*. In this case three human errors and four design errors contributed to the incident. In his conclusions, Mr. Epler states:

> The errors and failures cited are not individually unusual, although it would ordinarily be expected that they would be corrected during early operation and system shakedown. Engineered protection systems are not operated routinely and, as a consequence, error and failure modes can lie dormant and unsuspected, only to appear when emergency operation is required . . .
>
> The incident was not the result of a single failure but resulted, amazingly, from seven failures or errors in each of three identical channels, a total of 21 failures. If any one of these had not occurred, the reactor would not have been operated without emergency cooling. It is also noteworthy that this incident hap-

---

Ibid., p. 1622.

pened in a plant with an outstanding safety and availability record . . .*

Aside from the chance of a serious accident, these delays in safety research or its counterpart, proceeding too rapidly with the development of the nuclear energy program, may have forced us into the position where we shall have to accept far more risk for our electrical power than was necessary. Moreover, we may end up with a less reliable source of power. The reactors may have to be shut down frequently because of unforeseen engineering problems.

For example, in the May 14, 1970, issue of *Nucleonics Week* there is a fairly long discussion on the problems that developed with furnace-sensitized stainless steel in critical areas of the reactors. This article indicates that trouble was encountered at the reactors at Oyster Creek, Tarapur, Nine-Mile Point, and LaCrosse. These problems developed in furnace-sensitized stainless steel safe ends and other miscellaneous supports in the reactors.

A somewhat similar problem developed in the Indian Point reactor (May 20, 1970) where small pieces of material were found circulating in the cooling water. Since these reactors were constructed to meet critical power needs, it appears quite possible that brown-outs will occur when nuclear reactors fail. The possibility looms larger as we proceed to larger plants, each plant being a significant part of the energy supply.

---

*E. P. Epler, "The ORR Emergency Cooling Failure." *Nuclear Safety* 11 (4), 323-327, 1970.

Accidents In Fast-Breeder Reactors

The comments above concerning the water moderated reactors apply even more pertinently to the fast-breeders. Dr. Edward Teller expressed quite well the concern of many scientists and engineers, relative to fast breeders, when he wrote in *Nuclear News:*

> For the fast breeder to work in its steady-state breeding condition you probably need something like half a ton of plutonium. In order that it should work economically in a sufficiently big power-producing unit, it probably needs quite a bit more than one ton of plutonium. I do not like the hazard involved. I suggested that nuclear reactors are a blessing because they are clean. They *are* clean as long as they function as planned, but if they malfunction in a massive manner, which can happen in principle, they can release enough fission products to kill a tremendous number of people.
>
> . . . But, if you put together two tons of plutonium in a breeder, one tenth of one percent of this material could become critical.
>
> I have listened to hundreds of analyses of what course a nuclear accident can take. Although I believe it is possible to analyze the immediate consequences of an accident, I do not believe it is possible to analyze and foresee the secondary consequences. In an accident involving a plutonium reactor, a couple of tons of plutonium can melt. I don't think anybody can foresee where one or two or five percent of this plutonium will find itself and how it will get mixed with some other material. A small fraction of the original charge can become a great hazard.*

In his book, *The Careless Atom,* Sheldon Novick describes a number of accidents that have occurred

---

*Edward Teller, "Fast Reactors: Maybe." *Nuclear News* (August 21, 1967).

with nuclear reactors. One of these occurred at the Fermi Reactor, 30 miles from Detroit, Michigan. This is our first and only large scale fast breeder. In this accident some of the fuel rods had melted. The situation described above by Dr. Teller had occurred. Mr. Novick quotes Walter J. McCarthy, Jr., Assistant General Manager of the Power Reactor Development Company that owned the reactor, as stating that the possibility of a secondary and very serious accident was "a terrifying thought."

The terrifying thought involved the possibility of the melted fuel reassembling into a critical mass and resulting in an explosion that could lead to the consequences foretold by Dr. Teller. It was a month before careful attempts were begun to remove the damaged fuel elements. When nothing happened, everyone breathed a sigh of relief.

Dr. Teller says, "So far, we have been extremely lucky." But is Dr. Jordan's statement that the risk ". . . is certainly worth it" really true?

How safe and reliable are nuclear power reactors? Apparently, no one really knows. The United States is engaged in a gigantic experiment. The stakes which each individual must gamble in this experiment may be extremely high, possibly even his life.

CHAPTER 7

Nuclear Electricity And The Citizen's Rights

Every aspect of the determined public relations campaign of the U.S. Atomic Energy Commission, the Joint Committee on Atomic Energy, and the electric utility industry shows an infringement on the rights of U.S. citizens. The misuse of public funds for this purpose should raise the eyebrows of even the most cynical observer.

In the Declaration of Independence are the following historic words:

> "We hold these truths to be self-evident: that all men are created equal, that they are endowed by their Creator with certain inalienable Rights, that among these are Life, Liberty, and the Pursuit of Happiness . . ."

It is becoming increasingly clear that our democratic rights to the pursuit of happiness, in the form of a livable environment, are being seriously curtailed.

It is no secret that we face an environmental crisis of deep proportions because technological developments have resulted in massive pollution of our air, our land, our rivers, streams and oceans. It is also no secret that electric power generation is a major offender. Not only does the generation of electricity pollute in a serious and direct way; it also provides the power for a host of additional industries which pollute massively.

Platitudes abound from the electric utility industry and the U.S. Atomic Energy Commission concerning electric power "needs." Facts are curiously lacking.

Instead of a painstaking analysis of how increasing electric power delivery was being used, the dogma was advanced that electricity production *must* increase 10 percent per year, as it did for several years, far, far into the future. Projections of this, at least to the year 2000, are commonplace. This dogmatic projection, in the total absence of any rational examination represents a national disgrace.

Cliches such as "power means progress," or "we need more cheap electric power," or "growth is the cornerstone of civilization" are quite shopworn and overtly dangerous for the continued existence of life on earth. But no forum has been opened to consider the issue of optimum electric power production.

The electric utility industry and the Atomic Energy

The electric power industry promotes increased consumption of electricity through expensive ads like this one. The electric heating promoted here not only requires vast amounts of electricity; it is also more wasteful than any other form of heating. In order to produce 1 unit of electric heat, 2½ units of fossil fuel are burned. And in the process, 1½ units of heat are totally wasted.

Commission have been conducting a joint public relations campaign to sell the 10 percent annual growth in electric power production as a magical requirement of existence. And they pay for the campaign with public funds! This misuse of taxpayer funds by AEC is a scandal. The AEC admits doubling its public relations staff from 35 to 70 full-time Public Relations people, to "sell" the atom. Instead, the AEC and the electric utility industry should be sponsoring a serious public forum on the subject of electric power requirements.

Indirectly, the electric utility industry is using tax money to brainwash the public through ads in national magazines, TV spots, etc. What funds the utilities expend are regarded as part of their tax deductible "costs." The public pays for these *in addition* to the regular charges it pays to provide a profit for the utilities.

So two groups—the U.S. Atomic Energy Commission (and its Congressional Joint Committee patrons) and the electric utility industry—both promote their wares, with an apparent disregard for the public's right to understand, and participate in a meaningful debate and decision concerning electric power requirements. This represents blatant disfranchisement of the public — use of public funds for propaganda without any public participation.

The inalienable rights to life, liberty and pursuit of happiness are even more seriously infringed upon by the development of nuclear electricity in its rash proliferation. Very few citizens are aware of two major ways this comes about.

For many Americans, the purchase of a home is an important step in their pursuit of happiness. And because of the risk to that happiness, inherent in a loss of their home, Americans are accustomed to buying home insurance to protect that crucial investment of life savings. Little known to most Americans is the presence of a "Nuclear Exclusion" clause in their homeowner's insurance policy. A typical set of nuclear exclusion clauses from a Homeowners' policy issued by Hartford Insurance Group (one of the nation's largest and most reliable insurance companies) are as follows:

> 2. *Nuclear Clause—Section I:* The word "fire" in this policy or endorsements attached hereto is not intended to and does not embrace nuclear reaction or nuclear radiation or radioactive contamination, all whether controlled or uncontrolled, and loss by nuclear reaction or nuclear radiation or radioactive contamination is not intended to be and is not insured against by this policy or said endorsements, whether such loss be direct or indirect, proximate or remote, or be in whole or in part caused by, contributed to, or aggravated by "fire" or any other perils insured against by this policy or said endorsements; however, subject to the foregoing and all provisions of this policy, direct loss by "fire" resulting from nuclear reaction or nuclear radiation or radioactive contamination is insured against by this policy.

> 3. *Nuclear Exclusion—Section I:* This policy does not insure against loss by nuclear reaction or nuclear radiation or radioactive contamination, all whether controlled or uncontrolled, or due to any act or con-

dition incident to any of the foregoing, whether such loss be direct or indirect, proximate or remote, or be in part caused by, contributed to, or aggravated by any of the perils insured against by this policy; and nuclear reaction or nuclear radiation or radioactive contamination, all whether controlled or uncontrolled, is not "explosion" or "smoke." This clause applies to all perils insured against hereunder except the perils of fire and lightning, which are otherwise provided for in the nuclear clause contained above.

Many citizens are under the illusion that such exclusion clauses apply to nuclear war. Nothing could be further from the truth. If a nuclear electricity plant (or any of its necessary related activities, transport, fuel cleaning, or waste disposal) results in radioactive contamination of one's home, these nuclear exclusion clauses in homeowners' policies mean the citizen may lose the investment in his home, even though he has taken the wise precaution of buying insurance.

The astounded citizen might ask why the insurance industry sees fit to make a special exclusion of nuclear or radioactivity damage to his home. The insurance industry does not add a premium for coverage against radioactivity or nuclear damage. They just refuse to insure.

What nuclear or radioactivity damage worries the insurance companies? Is it nuclear war? Hardly. For if it were, they could readily so specify in the policy.

Clearly, the insurance industry, known for carefully protecting its profits, has taken very definite notice of the burgeoning nuclear electric power industry. It is obvious that it doesn't like what it sees *at all*. This lack of confidence in the safety of the nuclear electricity industry is expressed by the Nuclear Exclusion clauses in homeowners' policies. Underwriters refuse to risk dollars on the fail-safe formula developed for the nuclear electricity industry.

Considering the insurance industry's long history as a profit-maker, the public would be well advised to take heed of its extreme skepticism.

The insurance companies saw the nuclear electricity industry as a hazard, and moved *quickly* to protect themselves. The public is denied a similar opportunity.

The Constitutionally Questionable Price-Anderson Act

In the earliest days of the peaceful atom, there were wildly optimistic projections that electric power would become so inexpensive through nuclear electricity generation that metering the electricity would be hardly worthwhile. Those economic forecasts have proved sadly incorrect. In spite of massive subsidies by the federal government, direct and indirect, nuclear electricity is hardly holding its own against fossil-fueled electricity generation. And, it must be pointed out, the latter receives no federal assistance.

The Atomic Energy Establishment, embarrassed by its great promises and great expenditures, wanted to make *some* public showing that nuclear electricity generation was moving ahead, as advertised. But the leaders of the electric utility industry were disinclined to invest in nuclear power, lacking insurance coverage against possibly catastrophic nuclear accidents. The private insurance industry, feeling that the risk of accidents was unknown, would no more insure the industry against major nuclear accidents than it would the public.

The AEC sponsored one well-known study of the potential cost of a serious accident in nuclear electricity generation. The published results *(Report Wash-740, known as The Brookhaven Report),* which considered reactors only 1/5 the size of those currently being developed and planned, still concluded that a serious accident could produce monetary losses up to 7 billion dollars—over and above the injuries and losses of life!

There is no evidence that new reactor-developments have lessened the potential money losses to be faced with a major accident. Engineering developments may have cut the risk of certain accidents, but the larger capacity of the newer plants may have offset this. Indeed, no estimate has been made that excludes an even larger possible loss from the new, highly experimental, nuclear electricity reactors.

So the private insurance companies refused full coverage for nuclear electricity plants, and the electric utility industry would not risk construction and operation of nuclear electricity generating stations without insurance coverage. An impasse had arrived in the development of "the peaceful atom." Sensing that their major promotion was in jeopardy, the Joint Committee on Atomic Energy came forth with a fantastically bold solution.

A bill was proposed, known as the Price-Anderson Act, which simply eliminated individual liability in the event of a major accident in a nuclear electricity plant. Originally this act set 500 million dollars as the maximum liability for a single nuclear plant disaster (more recently extended to 560 million dollars). And, in addition, all but 60 million dollars of the insurance up to this limit was to be provided by the U.S. taxpayer. So if we consider the 7 billion dollar potential loss projected by the Brookhaven report, we note that private insurance carriers, in spite of governmental prodding, refused to cover more than one percent of the potential loss. This probably makes nuclear-electricity generation one of the least attractive insurance risks known.

The key point, over and above the lack of confidence of the insurance industry in nuclear electricity plants, is the utter disregard of personal rights the Price-Anderson Act represents for the average citizen. Since the maximum coverage is 560 million dollars per nuclear electricity accident, and since the damage can run to 7 billion dollars, in a serious accident, the individual might recover only 7 cents out of every dollar lost, assuming he is lucky enough to emerge from such an accident with his life.

The insurance industry will not suffer. The electric utility industry will not suffer. Through the generous manipulations of the U.S. Congress (prodded by the

161

Joint Committee), only the citizen will suffer—in the name of progress.

If the Price-Anderson Act were repealed, as assuredly it should be, it is extremely doubtful that any future nuclear electricity generating plants would be built above ground. Indeed, it is extremely doubtful that any electric utility company would be so foolhardy as to continue operation of nuclear electricity plants already built.

Electric utility propagandists, and atomic energy entrepreneurs, state that the extreme skepticism of the insurance industry shouldn't put anyone off. The insurance industry, they tell us, refuses to underwrite the risk simply because there is no prior "experience" upon which to base an estimate of the risk of major nuclear power plant accidents. Precisely.

But there is much more to it than this simple truth. The industry is saying, in a most persuasive manner, that they (the insurance industry) have *no confidence whatever* in the hopeful, optimistic safety calculations of nuclear electricity propagandists, certainly not enough confidence to risk dollars.

Another area of disfranchisement of citizens by the nuclear electricity industry must be clearly understood. The Atomic Energy Commission and the electric utility industry are well aware of the public's great skepticism concerning the safety of nuclear electric plants. So they resort to a form of public relations that might easily be construed as bribery.

For a variety of obvious economic reasons, power companies prefer to install their nuclear-electricity-generating plants as close as possible to the heart of major metropolitan centers. Such installations mean minimum transmission costs and losses in delivering power from production site to site of utilization. If they could get away with it, the utilities would place these plants directly in the major metropolitan centers. Indeed, *if* the nuclear plants were as safe as the propagandists claim, there would be *no reason not to do so.*

Realizing they are ill-prepared to answer questions that may be raised in such a large community, the utility companies shrewdly avoid these locations. There is little that can make installation of a nuclear-electric-power plant look attractive in a major city.

But, in dealing with the small community, located near a major metropolitan center, a workable promotion scheme is available to the electric utility industry, along with the probable absence of the sophisticated knowledge of the real hazards. This promotional scheme deserves careful examination, since it is used repeatedly to take advantage of millions of citizens.

A small community is chosen, generally less than 20,000 population, some 20-40 miles from a major metropolitan population center. Of course, anyone even mildly conversant with nuclear-accident hazards realizes that a major nuclear plant accident, that close, can easily endanger a million or more residents, in a major metropolitan center, through the spread of radioactive poisons.

"Go play in the park."

It's possible, you know. The grounds adjacent to nuclear power plants are safe and clean enough for children's playgrounds.

In fact, today, most nuclear power plants are places of education and enjoyment for thousands of adults and children.

equate nuclear fuel sources with nuclear explosions. This is the result of far more publ about bombs than about power-producing nuclear fuel.

The fact is, rigid safety precautions make nuclear industry in the United States and ab perhaps the safest industry in the history of technology. Before the go-ahead is ever give build a nuclear power plant, the Atomic Ene Commission requi................. l own .."

The AEC and the electric power industry spend millions

The "softening-up" begins with an advance guard of utility propagandists whose job it is to convince the officials of the small community and its Chamber of Commerce that jobs will be created. The odor of money flowing into a community works magic. And the citizens of a small community are mesmerized by the prospect of a reduction in their taxes, such taxes ostensibly to be paid instead by the nuclear electric plant. These economic "incentives" are hard to resist. Such attrac-

Visitors Center and grounds of the Connecticut Yankee Atomic Power station, Haddam

nuclear power

American to an average of 5 millirems of
ion a year. (A millirem is 1/1000 of a rem,
andard unit of measurement of the bio-
l effect of radiation.)

osmic rays expose us to another 30 milli-
This varies widely depending at what
ion we live. Just living on a hill exposes us
ore millirems than if we lived in a valley
et below.

- Natural radiation is in the earth. Radio-
materials in the soil and rock expose
average 20 milli

**"Why can't electricity be made like it
always has without using anything nuclear?"**

It can, and is. Right now, only 1% of the
electricity generated in this country is
produced by nuclear power plants. The other
99% comes from fossil fuel (coal, gas or oil)
or hydro (falling water) plants.

However, this ratio will have to change to
keep up with future no

In th

promoting the idea that atomic energy is safe and clean.

tive lures are accompanied by that classic blandish-
ment, "Nuclear Power Plants are Good Neighbors," a
homespun slogan designed to make one almost expect
the nuclear power plant to baby-sit, restore happiness
to broken homes, or play pinochle with the old folks.

In a recent example of such blatant gimmickry, a
group known as MEPP is perpetrating this scheme on
the small community of Ipswich, Massachusetts. The
major target for disfranchisement is Boston. The

MEPP group labels itself as dedicated "To Conserve Ecology." MEPP publishes a monthly, entitled "Plum Island Sounding News," distributed without charge to the residents of Ipswich. The "News" presents a blissful description of nuclear electricity's wonders, dismissing at the same time, any real consideration of the hazards such a plant presents. And the "News" proclaims in a full-page highlight:

Tax base Without Nuclear Plant . . . $66 per $1000.

Tax Base With Nuclear Plant . . . $24 per $1000.

Thus, through a set of economic enticements, perpetrated upon a community of less than 20,000 residents, plus a whitewash of nuclear hazards, a great city of millions can be placed in jeopardy, without having any opportunity to participate in the decision.

This devious approach is used repeatedly throughout the country, with minor variations. The net effect is that 90 percent of all United States citizens can be placed at risk, powerless to do anything about this antidemocratic procedure that somehow characterizes nuclear-electricity promotion. Such publications as the "Plum Island Sounding News" are designated "educational," so, of course, the taxpayer foots the bill for them.

We must consider the "educational" effort practiced by the U.S. Atomic Energy Commission in more detail. In a recent speech, Mr. Howard Brown, Assistant General Manager of the U.S. Atomic Energy Commission, gave a talk entitled "The AEC Goes Public: A Case

Study in Confrontation."* As everyone knows, the AEC has been spending taxpayer dollars to sell the public on the wonders of nuclear electricity for many years, while carefully glossing-over any adverse information about the hazards of nuclear electricity generation. Mr. Brown starts out with a lament:

"We've had a public information program for 20 years and a lot of effort has gone into it. For example, we've put out something like 10,000 press releases. We have a film library of some 11,000 prints. We've put out some 50 annual and semi-annual reports. We've made hundreds of speeches over the years, held scores of news conferences, and have circumnavigated the globe many times over spreading the gospel of the peaceful atom. Despite this, the message wasn't getting through. So, last Spring, the Commission decided to take a more direct, a more personalized approach.

"Since last March, the Commissioners and the staff have attended 39 public meetings on the environment, delivered some 22 speeches on the environment, and attended 10 Congressional Hearings, and submitted over 300 pages of testimony. We've prepared 66 articles on environment-related matters. Over 140,000 copies of our booklet, entitled Nuclear Power in the Environment, have been distributed. We've more than doubled our staff effort in headquarters from approxi-

*"The AEC Goes Public. A Case Study in Confrontation," Howard C. Brown, Jr. Delivered before the Atomic Industrial Forum's Topical Conference on Nuclear Public Information, Los Angeles, California, Feb. 11, 1970. Available through the Atomic Industrial Forum, New York, N.Y.

mately 35 man-years to over 70 man-years, not counting regulatory activities . . ."

In spite of this massive infusion of taxpayer dollars into propaganda, the public resistance to nuclear-electric-power generation has grown remarkably. In fact, the more propaganda the AEC puts out, the more public indignation rises, for obvious reasons. The public is far more intelligent than Mr. Brown realizes. Facing the specter of a technology that can potentially eliminate the continued existence of all living things on earth, the public is indeed interested to hear about atomic power, but what they want is hard information, not the "gospel of the peaceful atom."

The AEC wants to provide information—that is, one-sided information. When a nuclear plant is planned for a region, the AEC will gladly send speakers, all expenses paid by the U.S. taxpayer, to tell the residents of the area how perfectly safe nuclear-electricity generation is.

But, if those same residents want speakers to discuss the potential hazards of nuclear power generation, they must locate these speakers themselves and then pay for them out of their own funds. Is this what the AEC calls helping to present a "balanced" picture?

If there is anything the AEC cannot handle, it is an honest, open forum discussion of the hazards of nuclear electricity generation. Operating within the safe confines of its own public relations circuses, the AEC fares very well. It can slander critics, preach "the gospel," and whitewash all hazards. Recently we determined to find out whether the Atomic Energy Commission could stand the light of scrutiny by a jury of unbiased scientists. The following challenge was issued. (January 28, 1970)

A SCIENTIFIC CHALLENGE TO THE ATOMIC ENERGY COMMISSION STAFF CONCERNING THE CANCER + LEUKEMIA RISK FOR RADIATION

"Chairman Holifield, we urge you to nominate a jury of eminent persons, physicists, chemists, biologists, physicians, Nobel Prize Winners, or National Academy of Science members, or American Association for Advancement of Science members—none of whom have any atomic energy ax to grind. We urge you to serve as Chairman of a debate. Dr. Tamplin and I will debate each and every facet of the evidence concerning the serious hazard of Federal Radiation Council Guidelines against the entire AEC Staff plus anyone they can get from their 19-odd laboratories, singly, serially, or in any combination.

With their 20-year background on this problem and their large staff to draw on, they should be razor-sharp at a moment's notice. We are ready now. If there is any valid reason for questioning our submission to peers and for questioning our evidence, this eminent jury of peers will certainly determine so. If the debate before eminent peers is not held, then by default, we think the entire country and the world will know the answer without further question."

In spite of numerous repeat offers to the AEC for such an open-forum debate on these most crucial issues, the AEC remains in hiding. And yet, education of the public is supposed to be a major obligation of the Atomic Energy Commission.

Nuclear electricity, as can be seen from everything discussed here, is being promoted with an impressive disregard for citizens' rights. A total lack of candor characterizes proponents' presentation of the hazard considerations. Gimmicks are used to disfranchise the citizens of major metropolitan centers. Citizens stand to lose their property, without compensation, in the event of nuclear accidents—assuming they *are* lucky enough to preserve their lives!

CHAPTER 8

The Nuclear Legacy— Radioactive Wastes and Plutonium

In 1963, the United States, the Soviet Union, and a number of other nations signed a treaty banning nuclear weapons tests in the atmosphere. Up to that time, the U.S. and the U.S.S.R. had exploded nuclear fission-type bombs equal to some 300-400 megatons of TNT in the atmosphere. The radioactive fallout from these bombs contaminated the land, the waters, the vegetation, the animals, and man himself. Indeed, concern over the biological effects of this devastating fallout was the main reason for the atmospheric test ban treaty.

A single large nuclear-power plant (1000 megawatts electrical) produces as much radioactive material *in one year* as a 25 megaton atomic bomb can produce.

If the nuclear power plants now on order are built, they will produce 10 times as much radioactivity, each and every year, as was produced by *all* the atmospheric weapon tests before the treaty. By the year 2000, nuclear plants now planned will be producing 100 times as much radioactivity each year. Unfortunately, the evidence, to date, indicates that appreciable quantities of the radioactive material will find their way into the environment and thence into man.

Waste disposal practices

Fuel reprocessing plants take the spent fuel rods from reactors and reclaim the fissionable material, leaving behind tremendous quantities of radioactive waste. The AEC has several fuel reprocessing and waste storage and disposal sites. At the present time, there is one private fuel reprocessing plant and several private waste disposal sites.

The one commercial fuel reprocessing plant is the Nuclear Fuel Services plant in West Valley, New York. The safety regulations under which it operates are a travesty on the public health. Data on the radioactivity released from the West Valley plant, published by the U.S. Public Health Service,* indicate that any person eating as little as one pound of fish per week from

---

*July issue: N. I. Sax, Paul C. Lemon, Allen H. Benton, and Jack J. Gabay, "Radioecological surveillance of the waterways around a nuclear fuels reprocessing plant." *Radiological Health Data and Reports* 10:289-296, 1969. August issue: William J. Kelleher, "Environmental surveillance around a nuclear fuel reprocessing installation, 1965-1967." *Radiological Health Data and Reports* 10:329-339, 1969.

Cattaraugus Creek, where the plant's wastes are released, would be exposed to the guideline dosage of 0.17 rad. The plant is operating at only a fraction of its planned capacity! Moreover, in the August issue of Radiological Health and Data Reports, it is stated:

> . . . Suckers are taken from Cattaraugus Creek for food, especially in the springtime. In addition, there may be a practice of grinding up the flesh and bone to make fishburgers.*

Nuclear Fuel Services is also licensed by the AEC for waste "disposal." As a result, the company has a burial site on the West Valley compound. The July issue of *Radiological Health and Data Reports* shows that 10-15 percent of the radioactivity in the creek is coming from this so-called burial site.** Even if the reactors maintain low discharge rates, it appears, from the West Valley story, that these fuel reprocessing plants and waste burial sites may well bring our ground water and rivers to the limit of the regulations.

The West Valley story demonstrates the true attitude of the AEC toward its regulations and one of the reasons why it is so reluctant to make them more restrictive. In the AEC's 1969 publication, *The Nuclear Industry,* we read the following:

> Intermediate level liquid wastes is a term applicable only to radioactive liquids in a processing status

---

*William J. Kelleher, "Environmental surveillance around a nuclear fuel reprocessing installation, 1965-1967." *Radiological Health Data and Reports* 10:335, 1969.

**N. I. Sax, *et al.,* "Radioecological surveillance of the waterways around a nuclear fuels reprocessing plant." *Radiological Health Data and Reports* 10:294, 1969.

which must eventually be treated to produce a low level liquid waste (which can be released) and a high level waste concentrate (which must be isolated from the biosphere).

Low level liquid wastes are defined as those wastes which, after suitable treatment, can be discharged to the biosphere without exposing people to concentrations in excess of those permitted by AEC regulations.

Wastes generated in the cold or preirradiation phase of the fuel cycle (from the mine to the reactor), as well as wastes resulting from research laboratories and from medical and industrial applications of radioisotopes, are generally considered as low level or low hazard potential wastes.*

The West Valley plant is discharging "low level wastes" which the AEC considers as of "low hazard potential." A 5 to 50 percent increase in genetic disorders and deaths plus a 10 percent increase in cancer deaths appears to impress the AEC as a small hazard.

The AEC's own waste disposal practices are no better than those of the West Valley Plant. In 1965, at the request of the AEC, the Committee on Geologic Aspects of Radioactive Waste Disposal of the National Academy of Sciences-National Research Council made a final review of waste disposal practices at the Atomic Energy Commission's installations. The Committee submitted its report to the AEC in May 1966.

The AEC immediately suppressed the report, ignoring repeated requests to release it. Finally in 1970, in

___

*The Nuclear Industry, 1969. Report "published annually to present the AEC's assessment of the state of the nuclear industry . . ." Prepared by the U.S. Atomic Energy Commission. Washington, D.C., U.S. Government Printing Office, 1969, p. 252.

response to pressures applied by Senators Frank Church and Edmund Muskie, the report was made public.* As expected, it is critical of the AEC's waste storage and disposal practices.

The Committee report contained two important conclusions:

1) None of the existing AEC disposal installations is in a satisfactory geologic location.

2) Present practices of disposing of intermediate and low level liquid waste, and all types of solid waste, directly into the ground, will, in the long run, lead to serious fouling of man's environment.

The AEC has still not changed its practices nor relocated its installations. The West Valley story demonstrates that the time for taking the Committee's recommendations seriously has already passed. We have no adequate means of containing these low and intermediate level radioactive wastes, and the proliferation of nuclear reactors is only going to compound an already serious problem.

"Everything But the Squeal"

When an industry, like Armour, has abundant by-products, difficult or expensive to dispose of, it attempts to find a use (market) for the by-products. The nuclear

*National Academy of Sciences-National Research Council, Division of Earth Sciences, Committee on Geologic Aspects of Radioactive Waste Disposal, John E. Galley, Chairman. Report to the Division of Reactor Development and Technology, U.S. Atomic Energy Commission, May 1966. Unpublished.

industry is no exception, Again consider the 1969 report of the AEC, *The Nuclear Industry,* in which the following can be found:

Useful By-Products From Reactor Wastes

Fission products, such as strontium, cesium, and promethium, recovered during irradiated fuel processing operations, are already finding some useful commercial applications such as industrial thickness gauges, food irradiators, teletherapy units, as a power source in remote weather stations, etc. Others, such as xenon, krypton, rhodium, and palladium, are being considered for recovery because of their potential use in the electrical, jewelry, oil, and chemical industries. Possible markets for the expanded use of these materials in the near future offer many challenging opportunities.

Late in 1968, the AEC announced that the Richland Operations Office would seek expressions of interest from industry in the recovery of fission products rhodium, palladium, and technetium from the Hanford high level wastes. (See AEC Public Release L-252, dated October 31, 1968.) Considerable interest was indicated by several firms and one, PPG Industries, is exploring the possibilities of recovering these fission products by a proprietary process, using a sample of the Hanford waste.

Of particular interest in the by-product category is neptunium, which is used as the target material in the production of plutonium-238. It is possible that at some future date there will be a very large demand for Pu-238, for use as a power source in our space program and also there could be large demands for the artificial heart program if it is successful. General Electric is offering to recover neptunium (as well as uranium and plutonium) from irradiated nuclear fuel at its chemical reprocessing plant being constructed at

Morris, Illinois. Nuclear Fuel Services, Inc., New York State Atomic and Space Development Authority, and all the other companies with interests in the chemical reprocessing business are giving serious consideration to this and other isotopes for which a market and economic conditions justify recovery.*

There are about 100 private firms that produce radioisotopes or convert them into products for medicine, science, and industry. Total sales of these companies are estimated at $53 million annually, consisting of about $8 million in basic radioisotope materials, $16 million in radiochemicals, $25 million in radiopharmaceuticals, and $4 million in radiation sources. In addition, sales of devices, in which radioisotopes are employed, total about $40 million a year. If the sales of products produced by radiation processing, auxiliary materials, and services related to radioisotope and radiation uses are included, the total commercial activity in the United States is at a level of several hundred million dollars annually.**

A very large fraction of these radioactive byproducts will eventually find its way into the environment. In the process, some of these radioactive products will produce serious, immediate consequences.

Consider the case of a young Mexican boy who found a cobalt-60 source. The source, highly radioactive, looked to him like a metallic marble and he put it in his pocket. The radioactivity subsequently made him ill and his mother put him to bed. She put the "marble" in a drawer in the kitchen. As a consequence, both the mother and the boy's little sister became so ill

---

*The Nuclear Industry, 1969. Op. cit., pp. 266-267.
**Ibid., p. 22.

that the maternal grandmother came to care for them. In the end all four died.

Other such tragic examples are available. However, it is important to recognize that the unknown genetic consequences of introducing this radioactivity to man's environment will pale these examples by comparison.

One of the major legacies of the nuclear age is radioactive waste. Discussions concerning the disposal of it are misleading, because radioactive waste is not disposable! Guardianship of nuclear waste is a more meaningful concept.

We are producing waste products that must be maintained in isolation from the environment for a thousand years or more. The AEC does not appear to recognize this fact.

Plutonium: The Ultimate Hazard

The worldwide inventory of plutonium is man-made. It was virtually nonexistent in the earth's crust before the U.S. atomic bomb program was initiated. By far the major use of plutonium today is in the manufacture of nuclear bombs.

Plutonium has several nuclides, the most important being plutonium-239 (Pu-239) which is used in the manufacture of nuclear bombs. But Pu-239 is intended as the major reactor fuel of the future, through the development of the fast-breeder reactor. Its extremely long half-life, 24,000 years, will keep Pu-239's radioactivity undiminished much longer than the recorded history of modern man.

178

The cancer producing potential of plutonium is well known. An amount as small as one ten-millionth of an ounce injected under the skin of mice has caused cancer. A similar amount injected into the blood streams of dogs has produced bone cancers. However, it is the lung that is the most vulnerable to plutonium.

The vulnerability of the lung to plutonium exists because plutonium exposed to air ignites spontaneously. As it burns, it forms numerous tiny particles of plutonium dioxide. These particles are intensely radioactive. If inhaled, they are deposited in the deepest portions of the lung. There they remain, immobilized for hundreds of days, and during this time their radiation is able to affect the cancer-sensitive cells of the lung. The tissue around the particle is exposed to a very intense localized dose of radiation.

Our colleague, Donald Geesaman, has made an extensive analysis of the scientific data related to the hazard of these highly radioactive particles. His analysis pointed up a very sobering fact: The experimental data indicated that when small portions of tissue are exposed to extremely high dosages of radiation, cancer is an almost inevitable result. In other words, irradiation by plutonium oxide particles appears to represent a unique carcinogenic hazard. Somewhere between a few and a few hundred such particles would be enough to double an individual's chance of developing fatal lung cancer. An ounce of plutonium can form 10 trillion such particles.

Once in the air, these tiny particles remain suspended for long periods of time. Thus, plutonium released to the environment will represent a very long term and serious hazard. Plutonium fires have occurred at the Rocky Flats plutonium plant in Colorado that is operated by the Dow Chemical Company. Plutonium is measurable in the soil surrounding the plant. If the plutonium industry burgeons, we can expect more and more such contamination.

In a talk entitled, "Plutonium and Public Health," presented at the University of Colorado at Boulder on April 19, 1970, Donald Geesaman stated:

> Finally I would like to describe the problem in a larger context. By the year 2000, plutonium-239 has been conjectured to be a major energy source. Commercial production is projected at 30 tons per year by 1980, in excess of 100 tons per year by 2000. Plutonium contamination is not an academic question. Unless fusion reactor feasibility is demonstrated in the near future, the commitment will be made to liquid metal fast breeder reactors fueled by plutonium. Since fusion reactors are presently speculative, the decision for liquid metal fast breeders should be anticipated and plutonium should be considered as a major pollutant of remarkable toxicity and persistence. Considering the enormous economic inertia involved in the commitment it is imperative that public health aspects be carefully and honestly defined prior to active promotion of the industry. To live sanely with plutonium one must appreciate the potential magnitude of the risk, and be able to monitor against all significant hazards.
>
> An indeterminate amount of plutonium has gone off site at a major facility [the Dow Rocky Flats plant] 10 miles upwind from a metropolitan area [Denver, Colorado]. The loss was unnoticed. The

origin is somewhat speculative as is the ultimate deposition.

The health and safety of public and workers are protected by a set of standards for plutonium acknowledged to be meaningless.

Such things make a travesty of public health, and raise serious questions about a hurried acceptance of nuclear energy.

Although the carcinogenic hazard of plutonium in the environment is a serious problem, there is an even more serious problem associated with plutonium. It can be used to make atomic bombs. Even without the fast breeder program, considerable plutonium is produced in the present-day reactors. In fact, government purchase of the recovered plutonium is one of the price supports for the nuclear power industry.

The Safeguards Problem

U-235 and Pu-239 have been used to manufacture atomic bombs. Obtaining weapons-grade U-235 is very difficult because the U-235 has to be separated from its chemically identical and much more abundant relative, U-238. However, Pu-239 can be separated from its breeding material (U-238) by chemical means. The spent fuel elements from a present day reactor therefore contain, in a relatively easily extractable form, the primary ingredient for the manufacture of atomic bombs (enough to make several bombs).

With the spread of nuclear reactors and the eventual change to the fast breeder, plutonium will become as commonplace as heroin and even more profitable. A serious, an unsolved and probably unsolvable, problem is— how to prevent this plutonium from falling into criminal hands, where it can be used for blackmail and black-market enterprises?

A front page article in *The Wall Street Journal* of Thursday, June 18, 1968, stated:

> Scientists are raising a horrendous new possibility. It is far too easy, they say, for a crazed man, a revolutionary or a criminal to make an atomic bomb.
>
> "I've been worried about how easy it is to build bombs ever since I built my first one," says Theodore Taylor, a nuclear physicist who headed the Defense Department's atomic bomb design and testing program for seven years. He says the once-secret information needed to build nuclear bombs became available in unclassified literature several years ago. He especially recommends the World Book encyclopedia for its explanation of how a bomb works.

The December 1969 issue of *Nuclear News* reported on the Nuclear Safeguards symposium held at the Los Alamos Scientific Laboratory on October 27-30, 1969. The article stated:

> There was general agreement at the end of the symposium that, although there has been good progress made in safeguards technology, the world is still a long way from a foolproof system. In fact, some expressed doubts that this goal would ever be reached. AEC Commissioner Clarence E. Larson, keynote speaker at the symposium's banquet, identified himself with this group when he said: "From a practical standpoint, we may never solve all the problems, but we must collectively undertake to find solutions and to make use of safeguards practices."*

---

*"Time may be running out—safeguards warning sounded." *Nuclear News* (December), p. 16, 1969.

Later in the article, the comments made by Mr. C. Bellino of Wright, Long & Co., during a panel discussion are reported:

> Bellino stole the show. A leading expert on auditing procedures, Bellino serves also as a special investigator to the White House and the FBI. He told the audience that the subject of his treatise was "assessing the threat of highjacking by the Mafia." After humorously defining his terms, Bellino became quite serious. He pointed out that a letter was received recently on Capitol Hill stating that every trucking firm in a certain state, which he did not identify, was Mafia-owned and -controlled. He noted, too, that out of a secret list of 735 so-called Mafia members, 12 are or were owners of trucking firms; two are truck drivers and at least nine are or were union officials.
>
> Bellino believes it highly probable that, if some foreign tyrant offers a "deal," U.S. racketeers would be interested in it. A truck carrying uranium or plutonium could easily be highjacked. The theft could just as easily occur at a warehouse or dockside . . .*

Representative Craig Hosmer of the Joint Committee on Atomic Energy made the following comments before the 11th Annual Meeting of the Institute of Nuclear Materials Management in Gatlinburg, Tenn., on May 25, 1970:

> Earlier this year the Attorney General of the United States cited the Kennedy Airport cargo handling apparatus as being under the control of organized crime. The same can be said of many other key transportation elements of this country too. When and if SNM [special nuclear material] ever becomes an article of illicit commerce, the transportation element of the nuclear fuel cycle will become most vulnerable

---

*Ibid., p. 17.

183

to diversions. We'd better be cinching up in this area all along the way.**

. . . Many people, including myself, do not regard as very convincing the Dr. Goldfinger scenario where James Bond thwarts holding Miami hostage for a zillion dollar ransom under threat of blowing it up with a stolen H-bomb. Stealing a 1,000-pound top secret bomb isn't exactly easy.

But when you think not in terms of stealing whole bombs, but of diverting very small amounts of SNM at a time and of the possibility of a profitable black market developing, you get on more credible ground. Black markets already exist from all kinds of "hot" goods. They are quite flexible in taking on new product lines. If an SNM black market develops, the sale price to some country, individual, or organization—desperately wanting to make nuclear explosives—has been estimated as high as $100,000 [sic.—this should be $1,000,000] per kilogram.

A gram is 1/1000th of a kilogram and 1/1000th of $100,000 [sic.—should be $1,000,000] is $1,000. Liberating a half gram of plutonium at a time from the local fast breeder reactor fuel element factory might be so small an amount as to be relatively undetectable even by the best black boxes and the sharpest eyed inspectors.*

At the 20th Pugwash Conference on Science and World Affairs held September 9-15, 1970, Drs. Patricia J. Lindop and Joseph Rotblat of the United Kingdom stated:

Finally, when discussing the problems involved in the use of nuclear energy one must not forget about another and possibly even greater hazard: the possi-

---

**"Hosmer: some plain talk on safeguards." *Nuclear News* (July), p. 36, 1970.
Ibid., p. 37.

bility of clandestine acquisition by governments or groups of individuals of weapon-grade materials. This will become more and more difficult to avoid as the number and size of nuclear reactors increase. A very efficient system of controls is essential from the beginning. The IAEA [International Atomic Energy Agency] has not yet produced convincing evidence that they can tackle this problem, nor has the Agency been provided with the funds necessary for a project of this magnitude . . .

Plutonium was indeed aptly named: Plutonium— the element of the Lord of Hell. What kind of social responsibility exists within men who strongly advocate a drastic increase in the worldwide inventory of this element? There are currently acceptable alternatives for electrical power generation and the future holds out great promise for even better means of generating electric power. These men would seem to be possessed with a death wish that encompasses all of mankind. Shouldn't they be stopped?

CHAPTER 9

Alternatives
Available to Us

The map on the following page indicates the status of nuclear power in the United States as of December 31, 1978. The inset on the map indicates that the total generating capacity, of all reactors planned, will be some 201 million kilowatts. This will represent some 15 percent of the generating capacity of the United States in 1980. Quite obviously, there are alternatives available to us. One is simply to plan on using 15 percent less power in 1980. We shall discuss this important alternative later.

But we have another alternative that does not require any reduction in power consumption. Philip Sporn, retired President and later Consultant to American Electric Power Company, prepared a report for the

Joint Committee on Atomic Energy, which shows that the electrical power industry has ordered some 91-million kilowatts more generating power than the projected growth in demand will require.* If the 89 million kilowatts of nuclear power were eliminated now, the supply would match the projected growth nicely. In other words, it is reasonable to halt the growth of the nuclear power industry right now, to await its guaranteed safe development. And meanwhile, we can also consider the alternative sources for electric power.

The Fundamental Question: How Much Power?

All the nuclear critics we know deplore dirty, fossil-fuel-generating plants as much as, perhaps more than, they do nuclear plants. No one can deny the ill effects of the noxious gases that belch from the chimneys of these plants. (As we shall show, fossil fuel plants do not have to be dirty.) But noxious gases and radioactivity are not the only objectionable by-products of electric power production. There is waste heat — enough waste heat to change our ecology drastically if our projected power needs are real. Subsequently, public discussions must not be restricted to questions like "at what temperature shall the heated water from a given plant be discharged into the public water?" "How much radioactive waste shall be discharged into our

---

*Philip Sporn, "Developments in nuclear power economics, January 1968-December 1969." A report prepared for the Joint Committee on Atomic Energy, Congress of the United States, dated December 31, 1969. Unpublished.

NUCLEAR POWER PLANTS

(For the nuclear plant nearest you, consult Appendix IV.)

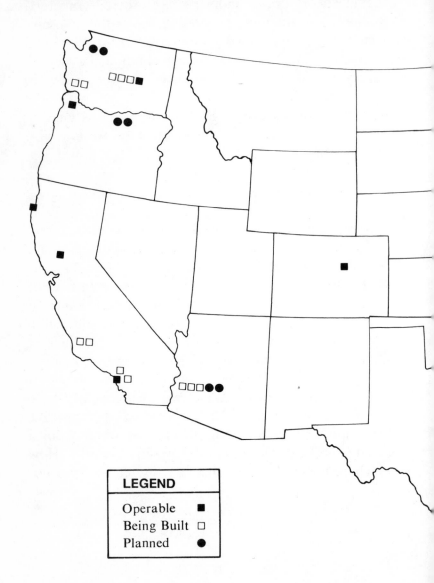

LEGEND

Operable	■
Being Built	□
Planned	●

N THE **UNITED STATES**

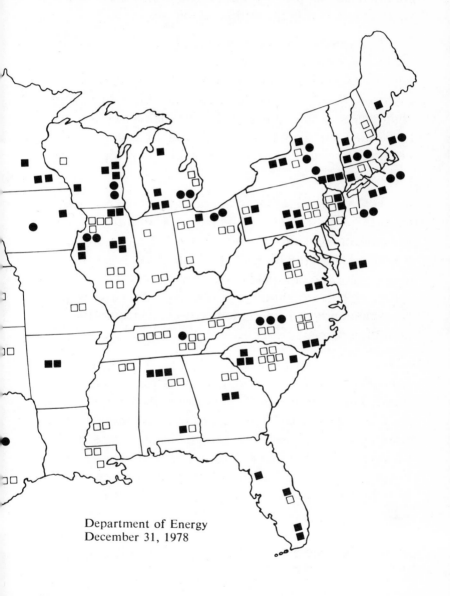

Department of Energy
December 31, 1978

common air supply?" To begin by asking these questions, is to begin in the middle of the story.

We must start with the fundamental question: *"Why more power?"* It is a question that has been publicly discussed only very recently. The flat, unqualified assertion that ". . . power needs are doubling every eight years" is not sufficient. Unqualified acceptance of this statement would be tantamount to endorsing the notion that electrical power consumption is a desirable end in itself.

Today, when environmental questions are paramount, we must question the basis for all intrusions on the environment. We do not know that more power is needed. The population of the United States grows by about one percent per year. It does not necessarily follow that a population increase of one percent per year demands an increased power consumption of about ten percent a year.

It is by no means certain that power asked for is equivalent to power needed. How is the power to be used? The advertisements by our utility friends stress the use of power for lighting hospital operating rooms, running audio-visual aid equipment in our schools, making possible stereo recordings of Brahms and Beethoven, and a host of other culturally interesting things. It is highly unlikely that these uses account for a significant fraction of the present or projected power use. Look closely, and it becomes readily apparent, for example, that the Pacific Northwest probably wants its added power to operate aluminum smelters in order to meet the growing need for beer cans and TV dinner trays.

That these factors are recognized, at the very highest levels of government, is evidenced by these excerpts from the keynote address at the American Power Conference in Chicago, given April 21, 1970, by Carl E. Bagge, Vice Chairman of the Federal Power Commission:

> Does it seem possible that it was but six years ago, in November 1964, that the Federal Power Commission, in cooperation with all the segments of this industry, published the first National Power Survey? This comprehensive nationwide survey was undertaken in order to define and articulate the long range goals of the industry. Some of the finest talent in government and industry studied the past performance of this highly fragmented industry; and as they observed the developing trends in generation and transmission; and as they projected the future supply and demand for electricity, there emerged a concept—a vision, if you will, which was translated into presumably attainable objectives—which were characterized as "guidelines for growth." . . .
>
> Looking back only the few years since its publication, one is struck by what in retrospect was an inexplicable lack of humility on the part of the architects of the National Power Survey. Certainty must have existed even then in the thinking of the utility industry and its regulators. The questioning of the limitations of technology, its direction, and even its values, which was then being focused on other sectors of our society, apparently had not extended to the electric power industry. And if it was, we must have

believed that the utility industry would remain immune from these forces.

How did this happen? How could we all have been so positive—so blindly certain—that the only challenge —the only goal—was the one which we conceived— —that of continually reducing costs in order to usher in the era of unlimited power—the era of the gigawatt —the electric energy economy—under what we characterized as "guidelines for growth." I submit that it was engendered by a monstrous sense of intellectual and technological arrogance which ignored not only the limitations of technology but even more importantly, the limitation of the vision of its high priests. The arrogance of our high priests is spread across the pages of our technical journals and in the National Power Survey as an irrevocable indictment of our own myopia. Today we stand convicted by our own testimony.

A little further in his talk, Commissioner Bagge stated:

Obviously, one of the most significant factors has been the sudden emergence of an almost religious fervor about the quality of our environment which has provided, within the political dynamics of this industry, a substitute for the old orthodoxy—the public's relentless demand for cheap power. Few issues have so captured the public's imagination. The speed with which it was transformed from a benign environmental ethic into a zealous ecologic faith has been nothing less than meteoric. Its sudden emergence as a national religion has profound implications to our theologians— and to the electric power industry. The environmental awakening will achieve even greater impetus throughout the Nation tomorrow when, to the accompaniment of teach-ins, marches, and demonstrations, hundreds of thousands of converts rivaling the crusades of Billy Graham will make commitments to the new religion of ecology.

And a little later he stated:

In the past, all of us have paid the price for the devastation inflicted upon the household of mankind by an industrial society. But it was assessed as a social cost which could be measured only to the extent that the benefits of our natural environment were denied to us. The damage was not reflected in the prices paid by us. Thus, the true cost of goods and services were understated to their competitive advantage. Today, we acknowledge that industry and consumers must bear the cost required to put an end to environmental degradation. This social cost must as a matter of national policy be recognized as a cost of doing business just as the cost of preventive maintenance reflects the price paid by the consumer for safety and reliability.

There is a tendency for the electrical power industry to equate *demand* for electricity with the *need* for electricity. On this subject Commissioner Bagge said:

The market-oriented philosophy reflected in your research effort has another outcropping in the form of promotional practices and promotional rates. While the industry has waged a campaign for an increasing share of the energy market, the success of these skirmishes has accelerated the already spiraling load forecasts and has created a level of demand which, in some cases, cannot now be met. It is paradoxical that the industry persisted in this objective at the very time there existed warning signs of forced load curtailments, through brownouts, voltage reductions and interruptions of service. In its quest for promoting greater electric use, this industry is now obliged to expend its resources to meet market demand—which it has itself, in part, created while experiencing difficulties in meeting normal market demand.

Commissioner Bagge is by no means alone in the critical questions he raises concerning electric power

consumption. Just recently the following article appeared in Time Magazine, December 28, 1970, under the title, "Heresy in Power."

"In 1967, when Charles F. Luce became Chairman of New York's huge Consolidated Edison Co., his first priority seemed clear. Since the average New Yorker then used only half as much electricity as the average American, Luce yearned to boost consumption,—and did. But last week he told a startled Manhattan audience: The wisdom of three years ago is the idiocy of today." Instead of trying to increase consumption, he now wants to decrease it.

Luce is regarded as one of the most socially responsible leaders in the utility business. He is also a realist. Crippled by equipment breakdowns, Con Ed has been forced to cut voltage in controlled "brownouts" for the past two summers. Meantime, New Yorkers demand even more power. Con Ed is all but helpless to supply it, because conservationists have won assorted court orders delaying the company's proposed new plants. They argue that power generation also generates pollution—and now Luce has publicly agreed with them.

As a long-term solution, Luce last week suggested a new federal excise tax of "perhaps one percent" on electric bills to speed new ways of generating power compatible with the environment. Until that luminous day comes, Luce is prepared to take an anti-growth position that other utility men might consider heresy. Urging New Yorkers to turn off unnecessary lights and appliances, he raises 'the serious question of whether we ought to be promoting any use of electricity'."

Mr. Luce is certainly to be complimented as a leader, in the electric utility industry, of the development of foresight that will be essential for this industry. Not only does he question the wisdom of stimulating electric power consumption, he recognizes the central importance of developing alternative methods for power generation—alternative methods with *minimal* intrusion upon the environment.

Further questioning of the wisdom of, or need for, major increases comes from the Editor of Science, Dr. Philip H. Abelson, (in *Science,* December 11, 1970, Volume 170, No. 3963) in an editorial entitled "Costs versus Benefits of Increased Electric Power." Especially important is his criticism of the wasteful use of electric power for home heating and of the misleading advertising which hides the pollution problem involved in electrical heating of homes. Dr. Abelson's remarks follow:

Costs versus Benefits of Increased Electric Power

Typical estimates of future demand for electric power in the United States assume a continuation of the previous rate of growth; power consumption eight times that of the present is projected for the year 2000. Little attention is devoted to the anatomy of the future demand. It is pointed out that population is growing, the gross national product is expanding, and energy demands are expected to increase. However, it is physically impossible for exponential growth to continue indefinitely. Already it is apparent that the generation and distribution of electricity entails some damage to the environment. Utilities can be expected to minimize the damage through the use of cleaner fuels, better siting, and underground transmission of power. However, some problems will persist. If conventional fuels are employed, the increased demands on them will speed exhaustion of oil and gas, and the use of large quantities of coal is likely to despoil large areas. Nuclear power carries with it many risks. Thus the utilities can expect to face continuing opposition in their efforts to expand power generation. The outcome of the battle is likely to rest on a balancing of social costs versus benefits to the consumer.

Much of the electric power goes to industry and to commercial use. However, the public is most immediately affected by that part going to individual consumers, and the electorate is likely to base many of its attitudes on personal experience.

If private consumers were to increase their use of power by a factor of 8 by the year 2000, where would the demand come from? Only a small fraction of the increase would come from population growth. There continues to be a proliferation of electrical gadgetry, but power consumption by most of these devices is trivial. For example, an electric razor consumes only a kilowatt hour per year, which is less than an air-conditioned house uses in an hour. In general, the devices that are used intermittently consume only modest amounts annually. Major items and their approximate typical annual consumption in kilowatt hours are color television, 500; lighting, 600; electric range, 1200; frost-free refrigerator-freezer, 1700; freezer, 1700; water heater, 3500; air conditioning, 5000; home heating, 20,000.

The more affluent segments of society already have about all the television sets, lighting, and cooling that they can use. Future expansion in public power consumption is dependent on an increased standard of living by the less affluent and on widespread adoption of electricity for home heating. At present only about 3.5 million homes are heated electrically; the major potential market is in home heating. Utilities are responding to the public's concern about pollution by extolling the virtues of clean heat. They soft-pedal the fact that the pollution problem is merely transferred elsewhere. However, it is technically much more feasible to eliminate pollution at a few major emitters than in millions of individual homes. Another consideration is the thermodynamic inefficiency introduced when electrical energy is dissipated resistively. However, if heat pumps were utilized at the homes, the overall efficiency would be acceptable. So-called all-electric living has a major disadvantage that should not be overlooked. It makes society terribly vulnerable to power failure, especially in winter.

The era of unquestioned exponential growth in electric power has come to an end. The future course of expansion will be determined by the public's estimate of costs versus benefits.

Professor Dean Abrahamson, of the University of Minnesota, in *Environmental Cost of Electric Power,* points out that the primary metals industry is by far the largest consumer of electricity.* The fastest growing and biggest user is the aluminum industry. It requires five times as much electrical power to produce aluminum as to produce steel. Substituting steel for aluminum, wherever possible, would conserve considerable electrical power.

Moreover, to reclaim steel from junk takes one quarter of the energy required to produce it from ore. *We should begin to recycle steel to the fullest extent.* One problem here is that copper contained in junked products makes it difficult to recycle steel. *Could automobile manufacturers be required to eliminate copper in the manufacture of cars?*

Aluminum junk is an indestructible litter. Recycling aluminum would conserve considerable electric

---

*Dean E. Abrahamson, *Environmental Cost of Electric Power.* A Scientists' Institute for Public Information Workbook. New York: S.I.P.I., 1970.

power. In Rodale's *Environment Action Bulletin* of July 18, 1970, the following appeared:

> On May 19 Reynolds Aluminum opened on aluminum reclamation center at 5455 Bonacker Drive in Tampa. What has happened since then is an example of what could be done if industry and the public would combine their efforts in an attempt to solve the country's solid waste disposal problem.
>
> Center manager James H. Riggs sums it up this way: "We didn't expect the terrific response from the public when we opened May 19. This is a temporary setup; we're planning a new plant a few blocks from here, and we may double the size of it." According to Riggs, when the news broke that Reynolds would pay 10 cents a pound (about a half-cent per can) for aluminum cans brought to the reclamation center, the plant was caught in a deluge of beer and soft drink cans—more than 10,000 tons during the first week of operation.
>
> Everybody appears to be getting into the act. For example, a retired man and his son came in with 760 pounds of cans. They'd picked them up in an eight-hour day along a quarter-mile stretch of road in Ocala National Park. A high school boy made $70 in one day bringing in cans. Even children come in with a few cans at a time, says Riggs.
>
> Reynolds, who has similar centers planned for New York, New Jersey, Houston, San Francisco and the Pacific Northwest, plans to get three percent of the total aluminum can production back.
>
> What makes the program work? It's profitable. Riggs explains: "To mine aluminum, you have to lay out millions for your initial capital investment; we've bypassed that outlay. To buy at $200 a ton is cheaper than mining it at $250 or $300 a ton.

This attractive difference in the cost per ton does not even include the substantial reduction in electrical power consumption. The aluminum companies can make an equal, or greater, profit from recycling the aluminum in all kinds of products, and at the same time the environment suffers less damage. What's wrong with that?

Unfortunately, power consumption is not synonymous with the standard of living. Power consumption relates more directly with the production of litter and solid waste and the decline in the quality of the environment. The biological evidence demonstrates that our technology became uncoupled from the standard of living in 1950. The increase in Gross National Product, since that time, has been associated mainly with the production of garbage.

The biological data demonstrate that infant mortality among fifty percent of the United States population (those below the median family income) is more than double that which occurs in the upper 25 percent income bracket; more than that, their life expectancy is reduced in excess of 8 years. For twenty percent of the United States population (those with the lowest family income) the infant mortality rate is four times that of the upper 25 percent income bracket, and life expectancy is reduced by more than 16 years. This situation has remained static since 1950. During the same period, environmental pollution has increased to near crisis proportions.

Our science and technology have not only become uncoupled from society; our environmental pollution and urban decay demonstrate that they are actually detrimental to society.

What we need is a master plan for improving the quality of life in this country. It cannot be a piecemeal operation, since each sector affects all others. We have the capability today to look at these large problems, to isolate the various interlocking factors, to determine the nature of the interplay, and to propose integrated solutions to the problems. We have the scientific and technological knowledge; we have the industrial capability within an extremely viable free enterprise system; and, as the space program demonstrated, we have a genius for organization. We could improve the quality of life in this country if we made the effort.

In his book, *Garbage As You Like It,* Jerome Goldstein shows a number of ways that we could do this by creating jobs with garbage while still turning a substantial profit and actually improving the environment.*

The present day industrialists, and many of their governmental counterparts, say that we cannot turn back the clock. No one concerned with the environmental crisis is talking about turning back the clock. Quite the contrary; environmentalists are concerned with ending the *laissez faire* line of least resistance approach pursued by our government and industry. We must demand that our government and industry leave this path, and challenge them to meet the real needs of

---

*Jerome Goldstein, *Garbage As You Like It,* Emmaus, Pa., Rodale Books, Inc., 1969.

201

society without degrading, but by improving, the environment. Our scientists and technologists should be urged to use their skills for solving the difficult problems that have been neglected for decades. Turn back the clock? Hardly! That is the problem; we are approaching the 21st century with 19th century concepts. The environmentalists want to set the clock to the correct time: Now. This was stated very eloquently by the noted scientist and philosopher, Bertrand Russell:

> . . . Whether men will be able to survive the changes of environment that their own skill has brought about is an open question. If the answer is in the affirmative, it will be known some day; if not, not. If the answer is to be in the affirmative, men will have to apply scientific ways of thinking to themselves and their institutions. They cannot continue to hope, as all politicians hitherto have, that in a world where everything has changed, the political and social habits of the eighteenth century can remain inviolate. Not only will men of science have to grapple with the sciences that deal with man, but—and this is a far more difficult matter—they will have to persuade the world to listen to what they have discovered. If they cannot succeed in this difficult enterprise, man will destroy himself by his halfway cleverness. I am told that, if he were out of the way, the future would lie with rats. I hope they will find it a pleasant world, but I am glad I shall not be there.*

Clean Fossil-Fuel Plants

In his talk cited earlier, Commissioner Bagge states:

---

*Bertrand Russell, "Science and human life." Quoted in *Scientists As Writers*, edited by James Harrison. Cambridge, Mass.: The M.I.T. Press. 1965, pp. 145-146.

Faced with these pessimistic developments in the nuclear field, it has become evident that the need for fossil fuel generation will appreciably increase. Even if nuclear generation should emerge as originally envisioned, fossil fuel generation will nearly double in the next twenty years. To many, especially those who had counted so heavily on nuclear generation, this realization has been slow and difficult. It even caught some segments of the industry unprepared. After all, who wanted to consider the environmental ugly duckling—when the promise of nuclear generation was ultimately to redeem this industry.

The problem posed by the necessity to rely on fossil fuel generation as the backbone of the industry for many years to come is compounded by the fact that low sulfur fossil fuels are simply not presently available in sufficient quantities to clear the air pollution hurdle which now has been imposed upon the industry . . .

Later in that speech, Commissioner Bagge explains the major reason for the air pollution from fossil fuel plants:

. . . The research and development effort for atomic energy received over 84 percent of all the federal funds for energy R&D.* It has also received approximately three billion dollars of government expenditures in the past twenty years. Compared with this ambitious federal commitment to atomic energy, the amounts of money which have and are being allocated for the improvement of fossil fuel generation and for other fossil fuel energy research are ridiculously small . . .

The September 14, 1970, issue of *Barron's* (national business and financial weekly) carried a front

---

*R&D means Research and Development.

203

page article, "Coal Imbroglio," that echoed Commissioner Bagge's assessment:

> . . . Legislation aside, Washington for years has subsidized the design, development and (through underpricing enriched uranium) operation of nuclear power plants. Official enthusiasm, generated largely by the Atomic Energy Commission, also succeeded in overselling the utilities on their state of perfection and ready capability, a miscalculation which the latter aren't likely to repeat. On the contrary, Duke Power reportedly has bought its own coal reserves, while Duquesne Light has agreed to finance the expansion of mine capacity . . .

> . . . Five years ago, if the AEC had had its way, coal might have been scuttled, and a temporary "crisis" mushroomed into a nationwide blackout. Private enterprise is vastly fallible, but it usually pays for and corrects its mistakes. The powers-that-be tend to perpetuate them or make them worse.

Thus we see that fossil fuels will form the backbone of our electrical generating capacity well into the 21st century. We can expect twice as many fossil fuel plants in the next two decades. So it is imperative that we develop pollution-free fossil fuel generating plants. In the August 28, 1970, issue of *Science,* Dr. Arthur M. Squires published a lead article entitled, "Clean Power from Coal."* In this article he demonstrates that we could have clean fossil fuel plants if we supplied the necessary research and development dollars, dollars that are now almost totally devoted to atomic power. Moreover, he demonstrates that the net result would be not only clean power, but *cheaper* power.

---

*Arthur M. Squires, "Clean power from coal." *Science* 169:821-828, 1970.

The power would be cheaper because the efficiency of the plants could be increased. As much as 50 percent more power could be produced per ton of fuel burned. This would substantially reduce the problems associated with waste heat.

Dr. Philip H. Abelson, editor for the American Association for the Advancement of Science, concluded an editorial in the September 25, 1970, issue of *Science* as follows:

> . . . In principle, all our energy needs could be met for a long time with coal. This raw material could be processed to yield sulfur-free fuel, liquid hydrocarbons, and methane. In practice, however, the development of the use of coal is limping along and is underfinanced. A few hundred million dollars a year devoted to research, development, and demonstration plants could be the most valuable expenditure the government could make.**

If we improved the efficiency of fossil fuel plants and made them pollution free, we could reduce the waste heat problem and literally create a breathing spell, wherein we can await the safe development of the power sources for the future.

Power Sources for the Future

Now that our declining environmental quality is beginning to receive proper notice, more and more of our best scientific and engineering talent is beginning to look at our means of generating power. We can therefore anticipate that some novel approaches will surface. One such approach is the use of the solar

---

**Philip H. Abelson, "Scarcity of Energy." *Science* 169:1267, 1970.

energy which bathes the earth each day. Although this idea has been denigrated for years, we should not cast it aside until these fresh looks at the problem have been given an opportunity.

Geothermal energy (underground steam) and tidal energy derserve imaginative consideration as potential additions to our pollution-free sources of power. Professor Robert Rex,* of the Institute of Geophysics, University of California, has recently reported that one geothermal field, running from the Salton Sea deep into Mexico, is massive. The steam available from just the U.S. portion of this reservoir of underground stored energy, he estimates, could provide 20,000 to 30,000 megawatts of power—as much power as the total current generating capacity of the entire state of California (27,700 megawatts).

Professor Rex estimates such sources of incredible amounts of power can last, undiminished, for 100 to 300 years. Numerous other resources experts continue to express enthusiasm for the pollution-free power, readily available to man, from such geothermal sources. A minor diversion of ill-advised funds in the AEC fast-breeder program could materially advance geothermal electric power development.

Energy-Efficiency: Our Largest Energy Supply

Fission power will never be acceptable as an energy source. And we have no need for it. We now know that the cheapest and largest source of energy available to us in the early future is energy-efficiency. We use our energy very

---

*Professor Robert Rex, quoted in "Pollution Free Power From Out of the Earth," by Marshall Schwartz, San Francisco Chronicle, (October 29, 1970).

inefficiently now; in fact, several countries in Western Europe achieve a standard of living equal to ours with only half as much energy-consumption.

In the U.S., we waste about 45 percent of the energy we use. We are not talking about changing life-styles, and we are not talking about doing without energy, either in our economy or in our daily lives. We are talking about how it lowers our standard of living to pay for energy and then to throw it away through inefficiency.

If we made simple alterations in the way we build our buildings from now on, and if we did some retrofitting of old buildings, we could save enough energy to substitute for the energy that would be generated if we built 430 giant 1,000-megawatt nuclear plants! To grasp how much additional energy that would be, remember that right now, the entire nuclear industry has only the equivalent of 50 large 1,000-megawatt nukes in operation (sometimes). It makes no sense to build hazardous and expensive nuclear plants in order to waste their energy! Simply by using energy *efficiently* in our buildings, we could create a supply of available energy far larger, far cheaper, far better for our standard of living and for employment, far more reliable for the economy, far safer than nuclear power—and ethically responsible.

There is an additional huge source of energy which we are right now *throwing away,* an energy source we once used until the electric utility industry destroyed it in order to increase its own business. That source is called "cogeneration" of power. Innumerable industries generate great quantities of steam for their industrial processes. If turbines were installed in many of those industries, they not only would generate their *own* electricity, but they could feed large quantities of surplus electricity into the country's power grids. Such electricity would be far cheaper than power generated by nuclear fission plants. Cogeneration of power this way could provide as much power as 200 giant 1,000-megawatt nukes.

So just these two kinds of energy-efficiency alone (in buildings and from cogeneration) could provide energy equivalent to 630 giant nuclear power plants, provide it far more cheaply, and help the economy and our standard of living at the same time. This we have learned since *Poisoned Power* was first published.

Naturally, the electric utility industry will do everything possible to prevent such constructive ideas from becoming reality. They may well create some *fake* power-shortages and brown-outs in order to frighten people into allowing continued operation and licensing of nuclear plants.

The hostility of the utilities to nonnuclear ideas has a simple explanation. While the average person may think that the utility companies would benefit by making electricity with less expensive installations, *exactly the opposite is true.* Through the collusion of public utility commissioners with the electric utilities, the ridiculous situation exists which allows the utility companies to make *more* money by building more expensive electrical generating plants. Utilities are "entitled" by such commissioners to make a comfortable percentage of profit on their investment. The bigger the investment, the bigger the profit. So the electric utility companies love nuclear plants *because* they are so expensive!

In the longer term, we must phase out the use of fossil fuels for generating electricity. Fortunately, in the years since

Poisoned Power was first published, it has become obvious that several ways of using solar energy are both technically and economically feasible. The amazing progress with solar energy technologies has been accomplished with pitifully little help from the government, since taxpayers' money is wasted so lavishly on nuclear power.

The statement that we must wait for the twenty-first century for practical solar power is an outright lie, sponsored and repeated by those who stand to gain by preventing solar energy from coming into widespread use. The great fear of the utilities is that solar power can be decentralized. The thought of losing control of our electricity supply, and of losing the monopoly prices which such control gives them, terrorizes the utility industry.

Between energy efficiency and solar energy, there will be an abundant supply of energy, the economy will be healthier, more jobs will be created, and our standard of living will be improved. *And* our dependence on OPEC oil will diminish dramatically.

CHAPTER 10

What Can Citizens Do About Nuclear Electricity?

**Recommendation for a Moratorium
On New Nuclear Electricity Plants**

It is undoubtedly clear to the reader by now that the authors of this book have grave reservations about the stampede to nuclear electricity being promoted through the AEC, the JCAE, and the electric utility industry. Wholly aside from the views of nuclear critics, a large, and steadily growing, citizen concern has developed. The question is repeatedly asked, "How can this madness be checked?"

Nuclear electricity generation, as it has proceeded up to now, is a classic example of the misuse of science and technology that has brought on our deepening environmental crisis. It is a particularly important case-

in-point because of the devastating possible consequences for all men and for all time. Once nuclear
pollution has occurred on a large scale—(and nuclear
electricity generation gives every promise of causing it)
there will be no hope of reversing the pollution for
hundreds or thousands of years.

Some gloomy individuals believe we might best let
the madness go forward, eliminate the human species,
and hope that at some dimly distant time in the future,
the biological accident that led to the development of
the human species in the first place, might occur again
with a better result.

Others believe a solution will come by another
route, short of obliteration of the human species.
Persons knowledgable in this field predict that a major
accident in nuclear electricity generation will occur as
a result of the proliferation of nuclear plants. They believe that before long such an accident will annihilate
the inhabitants of a major city, such as New York or
Chicago. If such a disaster should happen through a
nuclear accident, we would undoubtedly re-assess our
"need" for nuclear electricity generation.

It is a horrifying thought: — major calamity as the
route back to a rational approach to electric power
generation. However, the Atomic Energy Commission
and the Joint Committee on Atomic Energy are still
pouring taxpayers' millions into sales promotions de-
void of realistic safety considerations. Tragedy may
indeed be the ultimate resolution the problem finds.

Are there more reasonable solutions? People in a democratic society such as ours have been taught that the government's role is to protect our inalienable rights to life, liberty, and the pursuit of happiness. No wonder they are perplexed when virtually all branches of the national government operate in a way that seems to deepen the environmental crisis, rather than resolve it.

Numerous federal agencies operate in a way that either pollutes and destroys the environment directly, as does the U.S. Atomic Energy Commission, or they indulge the excesses of industrial pollutors. Congress may pass laws to protect the public — to protect the consumer. But somehow the very agencies Congress creates to provide the protection almost invariably end up subverting Congressional intent.

Congress put the regulatory and promotional (pollutional, really) agencies together in creating the Atomic Energy Commission. So, in this case, even the semblance of a separate governmental agency to protect the public's interest is non-existent. What can the public hope to accomplish by appealing to the Atomic Energy Commission for curbs on atomic energy?

The AEC allows for the public to be heard. The Commission announces formal public hearings prior to the issuance of construction permits and operating permits for nuclear power reactors. In principle, it is possible, therefore, for the public to intervene, to protest construction or operation of a reactor.

Who hears such protests? The Atomic Safety and

Licensing Board, selected from a panel appointed by the AEC. Nuclear power advocates dominate the membership of this panel. The chances for an unbiased hearing for an intervening group are clearly imperiled. It is true that some delays can be introduced into the whole system through such interventions, but, by and large, the delays are minimal. It is safe to say that if the public relies on open hearings in their fight against nuclear power plants, successive interventions on the next 600 nuclear power reactors will be followed by the construction and operation of the 600 reactors.

Nuclear electricity generation has developed under a set of, at best, questionable radiation standards — standards that are right now under sweeping review. Yet the licensing boards refuse to hear any challenge to a particular reactor that is based upon the invalidity and illegality of the radiation standards. The board *accepts* the standards as sacrosanct. It is up to the intervenor to prove that a proposed reactor will fail to meet the (current) standards. This is a patently ridiculous state of affairs.

This does not mean citizens should avoid such hearings. Certainly the opposing statements made there, reported in local papers, help educate the community on the true facts about nuclear power generation. Consider the effect: 500 people appear at such a hearing, all of them opposed to an impending nuclear power plant; this fact is reported locally. The community at large, and its officials, come to an early understanding of what they are up against, when the hearings produce nothing constructive and plans for the power plant go forward as if there had been no hearing.

Within the democratic process there are other avenues that can be effective. In the order of increasing effectiveness, these are:

(a) The U.S. Congress
(b) The State Legislatures
(c) Direct Public Referendum to Achieve a Moratorium on New Above-Ground Nuclear Power Plants.

Many members of the House of Representatives and the U.S. Senate are thoroughly informed on the true character of atomic energy promotion; if they are incensed, they feel powerless to do anything constructive. Largely the problem centers upon the stranglehold the Joint Committee on Atomic Energy has on the Congress. Suppose a bill were introduced into the Congress calling for a moratorium on construction of new nuclear electricity plants. The parliamentarian would undoubtedly refer it to the Joint Committee on Atomic Energy. There the bill will languish forever. Imagine what the chances are that the super-promotional Joint Committee on Atomic Energy would recommend a halt in such construction.

The early retirement of Congressmen Holifield and Hosmer would be helpful. More important would be action to keep all considerations of electric power generation, including nuclear power, from ever getting into the hands of this Committee. This does not appear imminent. An alternative approach would be to block

215

the annual Appropriations Bill for the Atomic Energy Commission, in an effort to force a reasoned consideration of nuclear electricity generation. All in all, it is difficult to develop much optimism about constructive action at the Congressional level, considering the archaic, obstructionist features of the existing Committee system.

It wouldn't do any harm, however, for a citizen to discuss these expedients with his congressman and senators. They might be willing to try some of the strategies outlined here.

It is extremely important to educate individual Congressmen and Senators concerning nuclear electricity generation and its hazards, not so much for what these men will accomplish in Washington, but for the influence they can have in their own states, where constructive action is definitely possible. And some effective measures might be achieved in the Congress itself, since the public is increasingly aware that politicians' lip-service, without action, only aggravates an already alarming environmental crisis. The early retirement of such Congressmen could change the complexion of the Congress enough to make progress toward a rational nuclear policy possible. But this would take time, possibly a few elections, and be too late to prevent much of the nuclear power plant proliferation currently being promoted.

The Fastest Way to a Moratorium

Individual state legislatures are awakening to the

concern of their constituencies, over the nuclear electricity juggernaut, especially where the plans call for nuclear electric stations that will leave almost no region of the state safe from the effects of an accident at one or another nuclear plant. Pennsylvania's state legislature recently responded to citizen pressure by initiation of extensive Hearings on Nuclear Electricity Generation, almost wholly focused on environmental, health and safety aspects of nuclear power.

Such hearings (in striking contrast with those conducted by the AEC) serve the extremely important function of providing the state legislators, and the public, with balanced information and an open-forum education on the less publicized aspects of nuclear electricity generation, such as health and safety. Until recently, the major source of information was the AEC's "gospel of the peaceful atom."

However, it appears that the most likely action of the state legislature will be to initiate interminable studies of the problem. Still, elected officials, when provided with full and honest information on both sides of a question, can aid materially in educating citizens of their own constituency.

To restore rationality to the nuclear electricity generation scene, the most likely avenue to success is a *moratorium* on new nuclear power plants above-ground for some period like 5 to 7 years. And the fastest way to achieve this is to get direct public vote, by initiative or referendum on the ballot, forbidding planning, constructing, or licensing such plants during the moratorium period.

The citizens of Eugene, Oregon were able to put a referendum on the ballot by citizen petition, which won a moratorium on construction of a nuclear plant that had already been approved. The action in Eugene proves that it is possible to educate the public about the dark side of nuclear electricity generation, in the face of a mountain of well-financed pronuclear propaganda.

It is very important for citizens to get involved, as directly as possible, in these major environmental issues. For too long, the public has been excluded from any significant participation in the dialogue, it has been thoroughly bypassed in considerations of what hazards to life and future will be accepted for specified benefits, or ostensible benefits.

This must change in the very early future. It is evident that the public is vitally concerned about the preserving of an environment habitable by humans and other living things. Public dismay at the progressive deterioration of the environment, with an almost total absence of any constructive action by government to alter this ominous trend, is equally evident.

Government agencies are often the chief promoters of pollution activity, aided and excused by huge public relations staffs that grind out reams of one-sided, uninformative press releases. The AEC is just one governmental agency which appears to have little regard for the public interest.

James Turner, a consultant with Ralph Nader's Center for the Study of Responsive Law, describes the incredible situation with respect to food additives thus:

"Over 95 different ingredients and chemicals can be added to bread by the manufacturer, as he desires, without adding them to the label. There are 76 such ingredients in soft drinks. There are 33 in cheese. *In fact the whole standard setting procedures of the Food and Drug Administration and the Department of Agriculture are irrational and do not reflect the best interest of the consuming public.*" (September, 1970)*

Michael Wollan** pointed out that an analogous situation exists in several other governmental regulating agencies, including National Science Foundation with respect to weather modification projects, the Federal Aviation Administration with respect to the SST project, and the Public Health Service with respect to fluoridation.

We have shown that the approach, philosophy, and methodology used in developing the radiation standards that govern the nuclear electricity program were erroneous. The electric utility industry has been misled concerning the radiation hazards associated with nuclear electricity. The leading physicist-engineers have been misled. Even the Chairman of the Joint Congressional Committee on Atomic Energy thinks we can safely be exposed to the amount of radiation it would take to produce a major public health disaster for this generation and for all future generations.

---

*San Francisco Sunday Examiner & Chronicle.
**"Controlling the Potential Hazards of Government Sponsored Technology." Michael Wollan, George Washington Law Review, Vol. 36, No. 5, pages 1105-1137, July 1968.

All this has been possible, and nuclear electricity generation has developed, as a major industry, through studiously maintained public ignorance of potential risks.

The only hope in this, and other desperately serious environmental problems, is to provide all the information to the public (all sectors of it). A growing segment of the public now realizes the hazards associated with nuclear electricity plants planned for above-ground construction on the very boundaries of major metropolitan centers. The full effects of catastrophic accidents that could occur remain unknown. Yet these gigantic, totally experimental plants are being constructed.

No responsible body of scientists, and no individual scientist, is *now* willing to minimize the potential radiation hazard to this and future generations. Indeed, when he was Secretary of Health, Education and Welfare, Robert Finch called for a total review of all radiation standards to reassess all the new evidence concerning hazards of ionizing radiation.

Those formulating the review are so concerned that they indicate it will take two years to study the evidence and to arrive at final conclusions as to just how serious these hazards are. Meanwhile, two Nobel Laureates* have come forward with their estimates, both even more severe than ours. A third Nobel Laureate** has

---

*Professor Linus Pauling and Professor Joshua Lederberg (cited earlier)
**Professor James D. Watson (cited earlier)

expressed himself opposed to nuclear power plant construction, simply because the gaps in our knowledge of radiation injury to humans are so large that he believes this alone justifies abandoning the construction of such plants.

There seems little doubt that the public must act directly to stop any further proliferation of this most dangerous, rashly promoted, nuclear electricity industry. The most constructive action seems to be a national moratorium on any further construction of nuclear electricity plants. In such a moratorium period, all the crucial questions can be thoroughly aired, and a rational future assessment planned.

It will be essential to press for action within individual states. In the 28 states for whom a nuclear electricity future has already been planned by those who never consulted the public, citizens groups are now thoroughly alerted and are seeking moratorium action in their own state.

In several states, the *initiative* is available and is the procedure of choice. Initiative is the procedure by which a specified number of voters may propose a statute, constitutional amendment or ordinance and compel a popular vote on its adoption. This is the highest form of effective participation in the democratic process. An initiative can be put on the ballot in such states to call for a 5-year or 10-year moratorium on nuclear power plant construction. Once passed, such a moratorium invalidates all attempts atomic energy interests might devise to have their way. Certainly where the initiative is available, it is the best course of action toward a moratorium.

The public must be prepared for a massive barrage of propaganda, from the Federal Atomic Energy Agencies and the electric utility industry. They will tell you that, after all, members of these bodies breathe the same air, drink the same water, and live on the same earth as the rest of us. And it is true. So true that if they can forget, for a moment, their immediate, parochial interests, they too might endorse a moratorium on nuclear power plants, at least until we can learn to handle this technology safely.

The ads about your "Good neighbor, nuke" will pour forth in the newspapers, the television, and in the lovely monthly utility company throwaways. Expensive 2-page newspaper ads will remind us that the western White House is located 4400 yards from the 430-megawatt San Onofre Nuclear Plant, and that it has not yet suffered injury. How can the American people, whatever their politics, stand idly by in such perilous times, while their President sojourns often at his Western White House? The region is one where a disaster evacuation plan is required by the AEC to allow operation of the nuclear reactor at all! One of the major accomplishments of a moratorium on nuclear reactors might be to remove the President of the United States from this senseless risk.

Whatever the pressure of the pro-nuclear, power-propaganda barrage, it is worth remembering that the American public is not stupid. Given an opportunity

In various parts of the country citizen groups have been formed to fight the construction of nuclear power plants. The above ad appeared in the local newspaper of Santa Cruz, California.

223

to look at the facts, they will surely decide *for* a moratorium on nuclear electric plants. Fortunately, the press has been reporting the grim prospects associated with radiation hazards fully and honestly, so the public is becoming informed. Further, the public is undoubtedly more interested in self-preservation than in preserving the AEC bureaucracy or the profit margin of the electric utility industry. The public is justifiably skeptical of an industry which proclaims that it must build new electric power plants to meet demands, then proceeds to spend millions for advertisements dedicated to increasing electric power consumption. The pollutors' cliches are rapidly becoming appreciated for what they are, a manifestation of total unconcern for the environment.

In the states where the mechanism of the initiative is, regrettably, not available, citizens should work hard to get it established, for the environmental struggle is only beginning. One has only to observe the politicians' inaction on environmental matters and the collusion of governmental "regulators" with those they regulate, to realize that traditional approaches will only hasten the deterioration of our environment.

A moratorium petition, signed by tens-of-thousands of constituents, can have a powerful effect in awakening sleepy state legislators and can even activate governors to take a position with the public on such matters. Those who refuse to wake up and bestir themselves may simply have to be retired. Obviously, the more names petitions contain, the more likely it is that state legislators will be shaken from their lethargy.

The informed part of the public, with respect to nuclear hazard, must become active in educating those who are still uninformed, or worse, misinformed. All the logic and all the evidence to counter the empty platitudes of the proponents of nuclear power are available. Proponents who are able to present any logical points to support their position in a debate are rare. Encourage the AEC and the utility representatives to debate in public forums, where any weaknesses and illogic in their arguments are exposed to full view. Man the negative side of the debate with the most knowledgeable combatants available. Each such debate guarantees additional support for the moratorium we are urging.

AEC and the electric utility industry suggest that, unless we go through with the nuclear power plans, the iron lungs in our hospitals will have to shut down for want of electricity. Such assertions can be countered with facts. It is industries that consume most electricity, not iron lungs, not stereo sets, not air-conditioners, and not electric toothbrushes. Recycling aluminum, for example, instead of wastefully using electricity to produce new aluminum, would help to solve our solid waste problem as well as our power problem.

The electric utility industry may one day find itself backing the effort to obtain a moratorium on nuclear electricity-generating plants. This industry is caught in a vice and is currently trying to extricate itself gracefully. Deceived by AEC assurances of cheap, safe sources of additional electric power, they invested prestige and billions of dollars in the nuclear enterprise. The directors of these corporations will realize, sooner or later, that they have made a disastrous mistake and decide not to throw good money after bad. The time will have come to cancel the "Good neighbor, nuke" propaganda line.

A moratorium on the building of above-ground nuclear electric plants can give us breathing space for some rational considerations of the power supply problem. If we make our determination to have it widely and firmly enough known, the research and development funds of the nation will go to those areas needed, to supply necessary, safe, clean electric power.

CHAPTER 11

Must We Hold Out For The "Cold Corpses"?

Proponents of nuclear electricity and other atomic energy developments are quick to claim that "we understand radiation hazards better than any other environmental pollution hazard." Another favorite: "We have proceeded with more consideration of safety in atomic energy than in any other industry." One yardstick used by these atomic energy enthusiasts is the heavy expenditure of public money in studying radiation hazards. A great deal of money has certainly been spent —much of it unwisely and inappropriately. As for the methods used to establish safe radiation standards for atomic energy development, less sound public health principles would be hard to imagine.

Here is what we foresee in public health losses if

atomic energy programs (including nuclear electricity) are allowed to proceed under our current allowable exposure standards—an average of 0.17 rad per year for Americans.

Cancer Plus Leukemia

Ten percent increase in the annual cancer plus leukemia death rate. One extra cancer for every ten occurring now. In the entire U.S. for 200-million people, this would mean 32,000 extra cancer plus leukemia deaths every year!

Genetic Diseases

Five percent to 50 percent increase in the rate of genetic changes. For 200-million people, this ultimately means 100,000 to 1,000,000 extra deaths per year from various genetic diseases, particularly heart attacks. If our population should ultimately grow to 300-million people, the genetic death toll would be 150,000 to 1,500,000 per year—not counting a five percent to a 50 percent increase in the incidence of socially crippling diseases as diabetes, schizophrenia and rheumatoid arthritis! (Contemplate a 50 percent increase in the major mental disease, schizophrenia; mental patients already occupy ½ of *all* hospital beds in the U.S.!)

All of these staggering projections in health cost are already accepted by numerous leading scientists worldwide. Some project precise cancer or leukemia figures at half as high, others say they will be three times as high. But the precise number is *not* the issue. The horrible realization is that the truth does lie in the tens of thousands of deaths annually, not the one case, or none, the Atomic Energy Commission suggests to the public. And though the projected genetic deaths are uncertain —between the huge numbers of 100,000 and 1,000,000 per year—we *are* certain the genetic cost will be staggering.

Surely some major flaw in logic characterized the entire approach to setting radiation standards if, in the 25th year of the atomic era, we find that the "safe" or "allowable" doses are so lethal. Actually, total illogic is the basic characteristic of radiation standards development, both for workers in atomic plants and for the population-at-large. And elementary reasoning shows us that, *if* we proceed to handle environmental poisons in the future the way we have handled the radio-activity problem up to now, our environment and our species are surely doomed.

Perhaps the simplest way to understand the erroneous approaches of the past is to ask how we might act if the problem were a new one.

Suppose we have just developed a new "wondrous" technology with a byproduct poison. For purposes of generalization, let us call this poison "Q". How much escaping "Q" would we be willing to tolerate in the environment where it might affect millions, or hundreds of millions, of humans? Of course, we must be concerned not only about whether people would drop dead immediately from exposure to "Q", but also about possible long-range effects upon individuals and upon the entire human species. Cancer and leukemia cases that might result in 5, 10, 15, 20 or 25 years must worry us. Genetic damage that might take generations to show up had certainly *better* worry us.

The promoters of the new technology would surely tell us (in two-page ads in all national magazines) that life on earth would be miserable unless the technology were immediately spread throughout the land. These same agents would probably wish to spend as little money as possible on protecting the people from exposure to "Q". Therefore, they would want minimal regulations against releasing "Q" into the environment.

How should society decide on the amount of "Q" that should be allowed to reach humans? Elementary logic would dictate that the promoters of the technology must prove safety of releasing *any* "Q" to the environment, that "Q" can do no harm to humans, *before* they release any "Q".

And how did we actually manage the question of radioactivity? The promoters of atomic energy and the bodies setting the standards said, in effect, the public must prove it is being harmed by radioactivity before we will stop radioactive pollution. Where environmental poisons are concerned, it has always been up to the public to show harm, rather than up to the pollutor to prove safety.

Should society say, with excellent reasons, that *no* "Q" should be released to the environment until its safety is established, it is certain to be faced with two of atomic energy's favorite cliches:

"Do you want to stop technological progress?"

"Don't you realize the benefits outweigh the risks?" Society answers, "Of course we wish to receive all the benefits technological progress can give us, but we insist on knowing the hazards involved. After all, we are the potential victims, you must convince us that what we stand to gain is greater than what we stand to lose. And if there is a risk, prove to us that we cannot receive the same benefits through some less hazardous means."

If the proponents of "Q" technology follow the pattern of the atomic energy promoters they will answer, "We just know the benefits are marvelous. The benefits just *must* outweigh the hazards. And furthermore, we have seen no evidence that the amount of 'Q' we plan to release will cause cancer, leukemia, and genetic damage to humans."

"But you are not saying that 'Q' has been proved safe," the public responds. "Your statement of 'no effects observed' simply reflects your ignorance concerning 'Q'. If you have made inadequate observations with 'Q', or none at all, how can you possibly know the answers?"

In answer, the "Q" promoters can be expected to appoint a body of expert scientists who will hold a long, serious conference and emerge from it with a magic number,—plucked out of thin air—a permissible standard for the safe release of "Q". And the public will be told it need have no fears, that the expert standard-setters will be watching the situation carefully. If too many corpses appear, they will confer once more, and set the safe standards for "Q" lower.

The public will certainly denounce the plan: "What utter nonsense it is to release the poison 'Q' into the environment and wait to see what happens! Surely there must be a more rational approach."

The "Q" technologists propose next that they be permitted to release "Q" in *some* amount. Presumably some accidental exposures will occur to sizeable groups of humans. The experts plan careful studies of how many cancers, leukemias and genetic mutations are occurring in the exposed humans. "Then we will know precisely how bad a poison 'Q' is. If the numbers should turn out too high, we'll reduce the 'permissible' levels of 'Q'." (This is precisely what happened with atomic energy. The standard-setters waited for the corpses to appear in Hiroshima survivors *before* they would believe increased cancer occurs in humans exposed to radiation.) Meanwhile, of course, all 200-million people in the country might have been irreversibly injured by the "Q" already released.

Obviously, *disaster* is the fate of a people willing to accept a poison in their environment, hoping that an accident will show them how dangerous the poison is. Worse yet, they will come to realize that technology spawns many "Q" poisons, not one, and all of them together might mean the end of life for the human species.

This entire scenario about the new poison, "Q," may sound far-fetched. But it is a precise description of how the radiation hazards question has been handled in the course of developing atomic energy. Far worse, both the nuclear electricity promoters and the standard-setting bodies still insist vehemently that they must be allowed to proceed with this same idiocy in the future.

Atomic technology was pushed hard by two governmental agencies, AEC and the Joint Committee on Atomic Energy. Accredited biological experts were assembled, in one committee or another, to consider radiation and radioactivity and decide how much people could be exposed to. Obviously, the pressure was on. These expert bodies must burden the atomic technology with the fewest possible restrictions.

Did these experts tell the technologists, "The burden of proof of safety is upon *you?*" Did they say, "We refuse to allow you to expose *anyone* to man-made radiation because we don't know how much physical damage it will cause?" They did not.

Instead, they pulled some numbers out of a hat and declared that the numbers represent "acceptable" standards for human radiation tolerances. And the atomic technology proceeded under the blanket of respectability of these "allowable" doses.

By now it is obvious, since these "acceptable" doses have had to be lowered 100-fold in the past two decades, and they have, something certainly was wrong with the original standards. Perhaps the experts *did* know that people wouldn't drop dead immediately from the "acceptable" doses they set at first. But for such late effects as leukemia, cancer and genetic diseases, the "experts" could hardly have been further off-base than they were.

If there had been no information available to the "experts" about the potential danger of cancer and genetic injury in humans, it might be argued that the men who set the standards had no way of knowing such radiation effects were possible. But the knowledge *was* available! These scientists *knew* that radiation causes cancer and genetic damage. And still, they set totally unacceptable standards! It is impossible to believe anything but that the agencies responded to pressure from the atomic technology promoters for "standards we can live with." The technologists were presented with a set of numbers for human exposure that presumably wouldn't make the promoters too unhappy, while those who set them probably prayed the disaster to the human species wouldn't be too severe.

The essence of this prayer comes through in the very forthright statement of ignorance made by the International Commission on Radiological Protection:

> (83) Because of the need for guidance in this regard, the Commission in its 1958 Recommendations suggested a provisional limit of 5 rems per generation for the genetic dose to the whole population, from all sources additional to natural background radiation and to medical exposures. The Commission believes that this level provides reasonable latitude for the expansion of atomic energy programs in the foreseeable future. It should be emphasized that the limit may not in fact represent a proper balance between possible harm and probable benefit, because of the

uncertainty in assessing the risks and the benefits that would justify the exposure.

It is very important to note that the International Commission says "this level provides reasonable latitude for the expansion of atomic energy programs in the foreseeable future." Is the concern for health, or for the technology? The Commission goes on to admit uncertainty *both* with respect to risks and benefits. It is almost unbelievable that an official standard recommending body would suggest allowing such exposure in the face of an overt admission of its own ignorance concerning hazards. But this is the record of such bodies, over and over again. The public must realize the implications.

If the errors in the earliest days of atomic technology are to be excused on the basis of ignorance or on the basis of a simple lack of awareness concerning sound public health principles, how shall we excuse the fact that rationality has not entered the picture to this date.

In our discussion of "Q," we expressed dismay that anyone might even *suggest* waiting for some catastrophic consequence of accidental human exposures to evaluate late effect hazards of the poison. Yet, this is *precisely* what the various standard-setting bodies for radiation exposure are doing and have been doing for many years.

Radiation Protection: Recommendations of the International Commission on Radiological Protection (Adopted September 17, 1965) ICRP Publication 9, Pergamon Press, Oxford, 1965.

A number of groups of humans actually have been exposed to ionizing radiation in high doses (Hiroshima-Nagasaki Atom Bomb Survivors, 14,000 British subjects with arthritis of the spine treated with x-rays). The experts have seized upon these groups with enthusiasm. They have pronounced that as the cancer and leukemia corpses appear in these human groups, they will be counted.

Only when a sufficient number of corpses have appeared, say, from lung cancer, will the experts accept that lung cancer is produced by radiation. If these men determine that too many cancers and leukemias are occurring, the allowable radiation dose to the public will *then* be lowered. Incredible as it may seem, this barbarian approach to public health practice is truly occurring!

Why is all this appalling? Suppose the accidental exposure at Hiroshima and in Britain had *not* occurred. There would have been no new information. Presumably, nothing at all would have been done about the allowable exposure levels which turned out to be so disastrous.

Now we know it takes 5 to 20 years before the various forms of leukemia and cancer show themselves after radiation exposure. If we sat by, waiting for each type of cancer to show up in the accidentally-exposed human groups, *decades* might go by, with hundreds of millions of people overexposed to radiation, and not a finger lifted to stop it. Exactly what is going on to this day!

Leukemia is the earliest cancer to occur following radiation exposure, surfacing after some five years. Obviously, before five years passed for the exposed human groups, it seemed that *no* cancers or leukemias had occurred among them. The standard-setters felt relieved. Then the five-year mark was passed, and leukemias did appear as an effect of radiation in the Japanese and British subjects. The standard-setters generously took due notice of the fact. Leukemia, they said, had indeed been produced in humans by radiation. Now the manuals were rewritten and leukemia was listed at last as a late hazard of radiation exposure.

What about the other forms of cancer that were known to follow radiation in experimental animals? The experts steadfastly refused to consider any of these. Not having seen any *human* corpses, they refused to admit they might exist, and would not lower the population exposure standards accordingly.

A little later, the incidence of thyroid cancer became so common in irradiated humans that it had to be acknowledged. The experts revised their reports of radiation hazards to include the thyroid cancer risk. They were now considering leukemia *and* thyroid cancer as radiation caused—but no other cancers.

So it continues up to the moment of writing this book. Even though cancer of the lung, the breast, the thyroid, the pharynx, the stomach, the lymph glands and bone have been unequivocally proved to occur in human subjects as a result of radiation, the standard-

237

setting bodies are just beginning to consider some of these cancers in their calculations.

Having embarked upon a course of action that can charitably be called public health-in-reverse, those charged with setting radiation standards persist in their errors and continue seriously to underestimate the true hazard of exposing people to radiation. And so nuclear electricity development and atomic technology in general both proceed under a set of standards permitting radiation exposure of the population that can lead to a massive public health calamity! All the while the public is reassured by announcements that eminent scientists are constantly reviewing the standards.

When we, the authors of this book, finally awakened to the unbelievable galaxy of errors represented in this standard-setting, we exposed it publicly. We were accused of making a direct, frontal attack on all radiation standards. Indeed, we *are* making a direct, frontal attack. And proudly. This account will, we hope, convince the public how long overdue such a massive, direct attack is!

Unfortunately, any criticism of erroneous public health practices is likely to be misinterpreted as an attack upon the motives of the men involved. We intend no such implications, nor even consider it in questioning their standard-setting procedures. These men are, after all, human. All of us learn through our errors, and few indeed have escaped serious errors of judgment in one or another aspect of their lives. But is it not tragic *and* inexcusable to persist in the errors of the past? The defensiveness of those scientists involved is leading directly to this tragedy. It appears certain that it will take public pressure to introduce a rational note into the radiation-nuclear energy scene.

We cannot refrain from addressing the issue of conflict-of-interest. And we do this not to impugn motives. Public officials are routinely required to divest themselves of holdings that might represent, or be considered to represent, a conflict with execution of their public duties. Yet, most of the scientists who serve on the various radiation standard-setting committees are directly or indirectly in the employ of the nuclear industry or the atomic energy government bureaucracy. Some are recipients of major university research grants from these same agencies.

The conflict of interest may be subconscious, but it is inescapable. Men can hardly be expected to consider civic responsibility exclusively, when they cannot be unaware that certain of their actions may well result in drying up sources of support for their research or for their salaries. This is a hopeless situation from which to extract objective performance. It is the very reason for our rather strict codes in such potential conflict-of-interest situations for public servants.

Recently one of us was lecturing in a university classroom concerning the leukemia and cancer hazard from ionizing radiation. A fellow professor attending the lecture asked, "If the Atomic Energy Commission pays to support your research, why do you criticize radiation as a hazard?" The deep implications of this question, undoubtedly asked in great innocence, must not be lost upon the public. If the source of research funds is expected to buy silence concerning hazards of major public concern, we are assuredly in very deep trouble as a society.

Many scientists would not ask this question so directly. They would simply remain silent about public-health hazards of technology if they sensed that speaking out might cost them their jobs or their research funds. Nor is it particularly hard to understand why. The heavy hand of reprisal by vested interests, governmental or private, is very widely appreciated.

Radiation Standards: How They Should Have Been Handled

We have noted that *everything* — the philosophy and execution for promulgating human radiation exposure standards—has been wrong. But this is the set of standards under which the nuclear electricity industry is proliferating. The conservative public health practice of caution on the side of health of the public, has been neglected—totally.

What would have been a reasonable approach? First and foremost, it is unthinkable to require human corpses before a standard-setting body will act to protect the public health. A procedure for effective action should have been developed, based on a cardinal set of public health principles, that does not require human experiments:

(1) At all times remember that our ignorance concerning biology and medicine is great, compared with our knowledge.

240

(2) In Ignorance, refrain.

(3) Where unknowns exist, *always* err on the side of
protecting the public health. Giving technology
leeway by *later* relaxation is always possible,
whereas alternate approaches can lead to irreversible human injury.

Why should the standard-setting bodies demand
human disaster as a guide for setting safe standards?
Experimental animal studies, available for decades,
prove conclusively what needed to be known. Virtually
every form of cancer and leukemia had already been
produced in several animal species provided radiation
was absorbed in the appropriate organ. Furthermore,
these studies indicated a five percent (or greater) increase in cancer occurrence rate for a variety of cancers, per rad of exposure.

A responsible society, applying sound public health
principles, would have assumed that *all* forms of human
cancer and leukemia would be induced *at least* as easily
by radiation as they were in the most sensitive experimental animal. Conservatism would suggest assuming
the human to be even more sensitive. Proceeding in this
manner, it would have been estimated that all forms of
cancer and leukemia would increase by five percent for
each rad of human exposure accumulated.

Now we can estimate how we would have evaluated the implications of a particular exposure level for
the population. Suppose we estimate the consequences
of developing nuclear electricity and other atomic programs working with the "permissible" dose of 0.17 rad
average for the population, the value chosen ten years

ago by the Federal Radiation Council. Now let us ask ourselves if this value, 0.17 rad, would really have become codified, or if the calculation would have led to far more stringent guidelines for radiation exposure.

It would have been reasonable to choose 30 years of age as a representative age for an "average" person who might be affected. By age 30, a person receiving 0.17 rad per year would have accumulated 30 x 0.17, or approximately 5 rads of total body exposure. At earlier ages, the accumulated exposure would, of course, have been lower. But evidence has long existed to indicate that the sensitivity to cancer induction by radiation is materially higher at early ages. (Now, we realize the sensitivity in utero is extremely high.) Furthermore, the lower accumulated dose at early ages would be counterbalanced by *more* than 5 rads accumulated at ages beyond 30 years.

By simple arithmetic, if 5 rads is the average accumulated dose, and if (as experimental animal data shows) there is a 5 per cent increase in cancer plus leukemia *per rad,* then, multiplying the two, we should have expected a 25 percent increase in the death rate *if* our population were exposed to 0.17 rad per year. This estimate would undoubtedly have shocked even the most hardened individuals. In the United States there are some 320,000 cancer-plus-leukemia deaths each year. A 25 percent increase would mean an additional 80,000 deaths from cancer and leukemia annually!

If this simple arithmetic had been done by the

standard-setting committees fifteen years ago, (the experimental animal data *were* available then) it is extremely doubtful that 0.17 rad per year would have been chosen as an allowable exposure. It is extremely doubtful that national programs like nuclear electricity generation would have been allowed to develop under such a guideline. The conclusion would have been self-evident; this is far too high a population exposure to contemplate.

The standard-setting committees did not go through this simple arithmetic. While their sincerity and devotion is *not* to be questioned; their judgment and comprehension of public health most certainly must be. The committees neglected the animal data which would have waved a red flag of alarm, and *demanded* human evidence before reducing the allowable radiation exposure for the public.

Now let us see whether this approach, using the experimental animal evidence, would have misled us or would have provided very sound guidance.

By now, to our sorrow, human evidence *is* available for all the major forms of cancer and leukemia induced by ionizing radiation. Whatever the results are for the few remaining minor forms of cancer, they cannot alter the picture significantly. Further, extensive human evidence shows a 2 percent increase in cancer per rad of exposure in young adults.

Again, using our 30-year old person as representative, with 5 rads accumulated at the allowable annual exposure, we have 5 x 2, or a 10 percent increase in cancer plus leukemia expected. And 10 percent of 320,000 is 32,000 extra deaths from cancer plus leukemia annually are to be expected *if* the population receives the allowable 0.17 rad per year.

Independently, Professor Linus Pauling estimates 96,000 extra cancer plus leukemia deaths annually. Comparing these estimates, 32,000 to 96,000 extra deaths annually, with the 80,000 that would have been arrived at from the experimental animal data, we realize immediately that the animal data would have provided sound guidance indeed! Moreover, the animal data would not have been at all super-conservative, for the human evidence now available shows there was no margin for safety!

The issue is *not* whether the estimate of 80,000 extra cancer-plus-leukemia deaths annually for exposure of the entire population at 0.17 rad would have been exactly correct. The real point is that the expected numbers would have been in the tens of thousands, *not* near zero.

Had this been appreciated, and announced fifteen years ago, nuclear electricity generation could have been more rationally evaluated in the light of realistic appraisal of the potential future hazard. The electric utility industry would not have been mistakenly lured into nuclear power by false and meaningless assurances of safety to humans in "allowable" doses of radiation.

It is truly pathetic to see how the misapplication of public health principles has deceived a major industry, including its executives, physicists, and engineers. The deep and widespread misinformation has led these physicists and engineers to design and install nuclear reactor systems under a delusion as to their true margin of safety. A real appreciation of the cancer-leukemia hazard of radiation would, doubtless, have altered the outlook of the nuclear electricity industry. Whenever design and engineering are carried through with a false idea of margin of safety, and in this instance false by 100 to 1,000-fold, real danger lies ahead.

We are not concerned with mistaken notions of the past, but of the present! Dr. Walter Jordan is a physicist, and Assistant Director of the Oak Ridge National Laboratory, a leading nuclear science and engineering laboratory.

Recently (May, 1970) in an article on nuclear electric power for the journal, "Physics Today," Dr. Jordan expressed his impatience with those who are concerned about the hazards of nuclear electricity generation. Dr. Jordan agreed there may be a hazard, but it surely is worth the risk. We would honor Dr. Jordan's privilege to express this opinion in any event. But we are horrified, upon reading his article, to learn that he has no concept at all concerning the cancer-leukemia risk! For Dr. Jordan states in his article that exposure to 30 times the allowable annual dose of 0.17 rad will lead to no physical effects upon the exposed individuals. Of course, Dr. Jordan cites *no* evidence to back his reassuring statement.

Let us explain to Dr. Jordan what a population exposure of 30 times the allowable dose would amount to in extra cancer-plus-leukemia deaths annually in the USA. Our estimate would be 960,000 extra deaths per year. Professor Pauling's estimate would be three times higher, or 2,880,000 extra deaths per year. Even a lower recent estimate, ascribed to R. H. Mole of the British Atomic Energy Authority, would lead to 210,000 extra deaths annually. Would Dr. Jordan consider that 210,000 to 2,880,000 extra deaths annually represent "no physical effect?"

Far more frightening is Dr. Jordan's recent appointment to the Atomic Energy Commission's "Atomic Safety and Licensing Board Panel." In this position, he will help review applications for the licensing of construction and operation of new nuclear-electricity generating plants. Here we have a man, obviously competent in his own field of physics and engineering, totally oblivious of the real hazards of radiation for humans. This man will be passing upon radiation safety and related matters for nuclear electricity installations. He will also sit on Public Hearing Boards to listen to any public protests and concern about the hazard of such plants.

Many nuclear reactor industry spokesmen and AEC officials have decried the "alarmism" associated with the estimates of leukemia and cancer risk from radiation, although there is not a shred of evidence they can offer in refutation of the estimates. It is *not* the estimates of the cancer and leukemia hazard from radiation that should alarm the public. But the public should be *extremely* alarmed that members of the

246

Atomic Safety and Licensing Board are totally oblivious to the real magnitude of the radiation hazard. As late as 1969, the Chairman of the Joint Committee on Atomic Energy expressed his opinion that there existed a margin of safety of 100 times in the allowable radiation dose for the public.

If the sound public health principles we have described had been applied, we could have averted today's sad state of affairs. Physicists, engineers, and utility executives could have been made aware of the true hazard of ionizing radiation. The rash proliferation of the nuclear electricity industry would surely not have occurred in the manner that it has.

The electric utility industry is a highly responsible one. It is a matter of great concern that it was so badly misled.

Considering the magnitude of the stakes involved, for the public, for industry, and the nation's future, it is imperative that sound public health practices be introduced into the nuclear electricity industry, especially since the hour is late for constructive action.

It is perfectly appropriate for a group of scientists, with expertise in a particular field, to provide estimates of the risk of serious disease as the result of potential exposure to environmental pollutants. In so doing, it is essential that the sound public health principles described above be applied in making the estimate of hazard. At all times a conservative approach, erring, where uncertainty exists, in favor of the public health is essential. Here the responsibility of the scientist, as a scientist, should end. The only appropriate standard for pollution is zero, until negotiated away from zero for very good reason. Expert scientists, operating behind closed doors, are in no way an appropriate body to make such negotiation, or to set any "standards."

The negotiation away from *zero* as the appropriate pollution level must necessarily involve a broad segment of society. For, in deciding to *allow* a negotiated pollution, society at large is accepting a hazard to health in current and future generations. Society as a whole must make the determination of whether the hazards are truly offset by projected benefits. The sole control over the health of humans and the quality of the environment can no longer be left in the hands of "experts." Such control must be carefully guarded and exercised by an informed public.

CHAPTER 12

Toward An Adversary System Of Scientific Inquiry

The recommendation of a moratorium on the construction of new nuclear electricity plants is directed toward elimination of a serious hazard to the human species. Of course, it is wise to avert any disaster once it is apparent. But it is relevant to ask why we must approach the brink of disaster so often, in the applications of technology. Nuclear electricity is only one case in point, though a profoundly important and dangerous one.

The public has every reason to ask why the nuclear electricity industry developed this far before there was a widespread appreciation of the hazards. Why, the public wants to know, was it not warned much earlier that the Insurance Industry has no confidence in nuclear electricity generation? How did it escape public notice that nuclear electricity plants represent a gigantic experiment being conducted at the peril of life and property of citizens of the U.S.? How does it happen that "standards" for radioactivity exposure (both for routine operations and in the event of accidents) are such as to lead to the expectation of massive injury in the form of cancer, leukemia, and genetic diseases?

The answers lie in the very nature of large-scale technology. One of its major characteristics is the careful exclusion of the public from all the considerations and decisions. Technologies, such as nuclear electricity generation, espouse the principle that, "In such complex problems we must put all of our faith in the *experts.*" The *experts,* for several obvious reasons, will surely bring society to its doom, unless certain corrective measures are urgently introduced. We shall consider such corrective measures in two areas:

    (a) the need for extensive public participation,
    (b) the need for adversary assessment of technology.

Technologies, such as nuclear electricity generation are highly financed enterprises, usually involving hundreds of millions, or even billions, of dollars. Biological scientists, physical scientists, and engineers are necessarily attracted to such technologies, because the research and development job opportunities are excellent.

The "experts" ultimately chosen to participate in decisions concerning safety, or lack of it, come from these same groups. They decide on "standards" for exposure of the public to such by-product poisons as radioactivity.

It is axiomatic: scientists chosen in this way are not likely to make decisions that embarrass their technology. And adverse decisions concerning its hazards can compromise the technology. A "standard-setting" decision that can make the technology itself appear economically unattractive might wipe out a scientist's financial support. Consciously and subconsciously, the scientist has a strong motivation to make the technology look good. The result, in general, is that the public bears the burden of any hazards, actual or potential.

Such scientists and engineers are not evil in their intentions. However, they are often so thoroughly compromised in outlook that their search for hazards can best be characterized by minimum, sincere diligence. At every step in their deliberations, where they must choose, the choice is that which minimizes the hazard estimate. Precisely the opposite choice should be the case if public health and safety were truly of paramount concern.

One product of such scientific deliberations is the concept of an "allowable," or "tolerable," or "permissible" dose of a poison such as radioactivity. Never has *anyone* proved that any dose of radioactive poison is safe. Yet bodies of scientific "experts" are duly appointed to "standard-setting" boards or committees. Under the auspicious title of "Radiation Protection," such committees proceed to ordain how much radioactive poison the public must accept in order to allow for "the orderly development of the technology (atomic or other)."

In the course of their deliberations these committees repeatedly recite the benefits of the new technology and state that society can ill-afford to forego them. Next they estimate the hazards, with all uncertainties weighted for the technology, not the public health, stating all the time that they are proceeding cautiously and conservatively.

As an early constructive step, the public could insist upon the abolition of *all* "standard-setting" bodies. Major decisions concerning exposure of the public to poisons such as radioactivity or other poisonous technological by-products *belong in the public forum.* Such decisions, often dealing with effects upon the heredity of the human species, are what we choose to call decisions for all men for all time. A very broad representation of society as a whole *must* assume active participation in such decisions.

How could such a broad segment of society make sound decisions concerning exposure to a poison such as radioactivity? There are several prerequisites:

(1) Abolition of "experts" or "standard-setters" as decision makers.
(2) Honest presentations of the hazards of by-product poisons.
(3) Honest presentations of the benefits of proposed

technologies, including serious consideration of alternative methods of achieving the benefits.

(4) Open-forum debate, followed by decision either by public vote or vote of *public* representatives.

(5) Preservation of the option to reverse decisions. New information concerning hazards and benefits must always be anticipated. Society must preserve the option to change its choice of technologies in the light of new evidence.

(6) Recognition of the principle that the appropriate permissible dose of a man-made poison is *zero*. Deviations from zero allowable pollution must be allowed only by public decision to be polluted in exchange for some benefit it chooses to receive.

(7) Recognition that the burden of proof is upon the technology to prove safety, rather than for the public to prove hazard.

Clearly, the major inputs are (2) and (3), the *honest* presentations of hazards and benefits. It is to be expected that enthusiastic supporters of the technology will be abundant, simply because dollars are associated with the technology. These proponents will describe the benefits glowingly; they will discover the hazards to be minimal or zero. Further, they will find alternatives to their technology to be non-existent or hopelessly difficult.

This all describes the nuclear electricity industry perfectly. It is what we can expect for just about any hazardous technology. And this can hardly be described as the kind of balanced presentation required for open-forum decision-making by the public or its representatives.

The obvious requirement is an assessment of benefits and hazards by competent scientists and engineers who *do NOT* derive their income and support from the technological entrepreneurs, private or governmental. What is needed, therefore, is an *adversary* system of technology evaluation. Such adversaries must provide the information the technological proponents might fail to provide. The public may be surprised to realize that this essential adversary evaluation of technology is totally lacking in our society.

The heavy hand of economic and job reprisal is so well appreciated by scientists and engineers that few actually involved in the technology will speak out against it. We must create a reprisal-free system of adversary assessment. We must learn how to fund such a system so that it cannot be silenced or inhibited by the entrepreneurs or their bedfellows in government.

Strangely enough, such an adversary system would cost very little. If it were mandatory that a few percent of the dollars that go into a new technology go into the funding of technology assessment, the resultant development of sound criticism of technology should be phenomenal. This would give the public a chance for a reasonable, open-forum debate concerning vital new technological directions.

Of course, the sponsors of up and coming technologies will, at first, regard it as a thwart. However, with more sober consideration, they may very well become major supporters of adversary assessment early in the development of a new enterprise. Unpleasant facts about a technology have a way of ultimately becoming obvious to everyone. The economic costs of realizing them too late can be extremely high.

Some may say that even in a reprisal-free atmosphere, scientists expected to do adversary assessment of technology might still be co-opted. This is a hazard, of course. On the other hand, there is a growing group of humans who do truly care about preservation of the human species and a livable environment. Such individuals could make a unique, effective contribution in the role of adversaries in the evaluation of new technologies. The dollar cost of establishing such technology assessment is trivial. The potential benefit for the survival of humans is incalculable.

Industry has long understood the danger of "yes-men" in high places. A technology, under current circumstances, is practically guaranteed to find itself burdened with a group of "think-alikes" throughout its technical staff, for the simple reason that those who speak out are shortly weeded out. This dangerous situation operates against solutions of major problems both for industry and for society. All facetious quips aside, it is unquestionably true that industry and society must breathe the same air, drink the same water, and share the same earth. Over the long pull, industry cannot possibly survive and prosper by conducting anti-human activities.

The problems presented by technology may be difficult, but they *must* be solved. A real dialogue, with opposing views placed in the open forum, represents the most constructive approach in working toward solutions.

Today we have only a monologue, in the absence of adequately supported adversary technological assessment. The early establishment of reprisal-free, fully funded centers for adversary criticism of technology can correct this serious situation, to the advantage both of enterprise and society as a whole.

CHAPTER 13

The Ultimate Issue– Conversion Or Ecocide

Whether the issue be consumer products—adulterated, falsely labelled, or unlabelled, with respect to potential or known toxic materials — or major technological projects, spewing long-persistent toxic pollutants into the environment, the ultimate issues are a livable environment, good health, and a decent quality of human life. That a considerable segment of our industrial-manufacturing-technological activities is seriously uncoupled from these goals is a truism.

Many are hopeful that by pleading, by exposure, by legal harassment, by public education, and by dedication in the public interest, we shall be able to turn all of this around, and thereby have technology finally begin to serve societal needs and goals. But we might, through focus on details of the injustices and reprisals, indeed win important battles, but lose the war to prevent ecocide. Is there some central theme that underlies all these problem areas, with features that militate strongly against local, isolated solutions?

We know the shortsighted parochial view of our economics, which fails totally to consider the health costs to society and the environment's deterioration.

We know the futility to date of the efforts to alleviate these burdens upon our health and future through governmental regulatory bodies. At best, this has produced no real relief, and is not likely to do so.

We know that the technologist or scientist who speaks out from an industrial or governmental position will certainly meet reprisals, censorship, and, most likely, unemployment.

Why don't we face squarely the real root of all our problems and ask ourselves whether a realistic, non-utopian solution is possible?

The Promotional-Profit Incentive

Our society is based upon the premise that initiative, innovation, and promotion, all leading to economic profit, will by their very nature insure the delivery of goods and services that will steadily upgrade the quality of life for the greatest number. The present environmental crisis clearly indicates that such a desirable result is anything but automatic. The threats posed by food adulteration, poisonous chemicals of agriculture and commerce, and radioactivity may, by synergistic activity, guarantee ecocide, with little or no opportunity for us to understand the hazard or to take remedial measures. Indeed, a quest for remedial measures for specific abuses may divert us from effective, broad action.

We do not think it is particularly meritorious to question the promotional or profit incentive. They are deeply ingrained powerful human motivations. Moreover, it appears that societies which have ostensibly eschewed the profit motive seem as capable of misdirecting technology as we are. And it may well be that desirable innovation should be abundantly encouraged. There is no doubt that skill and inventiveness should enable technology to operate in society's behalf and to provide many desirable and essential innovations, especially until a rational solution to our population problem is found—some time off at best.

No fundamental law exists, so far as we know, which dictates that a profit-oriented society must *necessarily* engage in anti-societal, eco-destructive pursuits. No fundamental law says it is impossible to make money doing worthwhile things. We may well exhort industry and technology to develop a sense of public interest responsibility, even to pinpoint the fact that a parochial view of their interests will destroy them along with the rest of society.

Such exhortations are justified, carry a real ring of

morality, and are by no means scaremongering or doomsday prophecies. It seems to us that they will fail, however, because they don't address the real problem. It is one thing to point out wrong directions; quite another to provide a realistic framework for effective solutions.

To come up with such solutions we must understand some powerful factors which characterize innovative, profit-oriented enterprise:

(1) *The investment of capital* by the entrepreneur-innovator. Today innovation and technology are very big business, most endeavors of any consequence encompassing in a short while the effort to distribute goods or services of the particular technology to 200-million people nationally, and to even larger numbers when foreign outlets are considered. Even the early investment is generally very large. If the particular technological entrepreneurial project has gone along for a period of time, the investment of capital funds soon becomes huge, and indeed a matter of considerable importance concerning which the entrepreneur must be extremely protective. It is a characteristic of innovation that there must be initial enthusiasm and promise—and this characteristic makes it very difficult to appreciate the adverse by-product effects, such as hazard to life. Two features operate here:

(a) The subconscious desire to look the other way for an innovation that holds promise of real utility and profitability.

(b) The widespread delusion that science and technology will undoubtedly provide a "fix" for any hazard of the enterprise.

(2) *The investment of career* by a large body of scientists and technologists who prepare themselves at great cost for the particular enterprise. And if the technology has persisted for any length of time, such men have achieved position, prestige, and a high personal economic stake in the future of the enterprise.

A case in point is the nuclear energy technology. Whole university departments have devoted themselves to the training of nuclear engineers and related technologists. And beyond the educational level, there are thousands of nuclear engineers, health physicists, and biomedical scientists with well-established careers predicated upon the continuation and growth of nuclear energy technology, in particular nuclear electricity generation. And this doesn't begin to take into account some 140,000 atomic industrial workers with a large stake in the continuation and growth of this industry. Indeed, the governmental regulators themselves have a *not* inconsiderable stake in the nuclear energy enterprise.

(3) *The investment of ego and prestige* by the elite who have thoroughly committed themselves to the glowing promises of the technology, in full public view. Again, the longer the enterprise has persisted before adverse features become evident the greater the ego-prestige commitment of such elite, and the more difficult it becomes for such elite to reverse their positions.

In nuclear energy, can any fail to understand the difficult position of Chairman Glenn Seaborg who has admitted his position as a prime salesman for nuclear electricity generation? From a myriad of platforms, and in countless printed statements, he has stated that "the atom came to us in the nick of time." Is anyone so naive as to fail to understand why Dr. Seaborg is having difficulty facing the realization that the hazard of ionizing radiation is far greater—20 to 30 times greater—than was thought a decade ago? Or to fail to understand why Dr. Seaborg dodges the question of the likelihood of a catastrophic accident at a nuclear power plant? Or to fail to understand why Congressman Chet Holifield, having pushed appropriations of billions for nuclear energy development through Congress, clings to the concept of a "safe" amount of radiation exposure—a concept rejected by a whole series of distinguished scientists, as well as all the scientific bodies involved in study of radiation hazards?

It should be unrealistic for any of us to hope that dangerously misguided technological-industrial endeavors will come to an end through:

> Economic suicide of the capital-investing entrepreneur,
> Career and job suicide of the technologists and workers,
> Ego and prestige suicide by leaders, promoters, or apologists for the enterprise.

To argue that a higher morality should guide all these men, with their varied, vested interests, is simply to produce a totally unreal and unuseful image of men. It is obvious that long-range *ecocide* will necessarily win out over short-range, parochial economic suicide, career-suicide, or ego, prestige-suicide. And morality won't even visibly enter into the consideration, for the mechanisms of rationalization will surface in abundance to protect against even the most obviously indefensible position.

Limited Victories

Some may point out that, in spite of all the above, we can win the battle in the existing framework. The battles, yes; the war, no. Cyclamates, it will be argued, have been withdrawn from the market in spite of vested producer interests, in spite of shenanigans of the most reprehensible character from the Food and Drug Administration. But for every cyclamate withdrawn, there are hundreds or thousands of compounds in the food additive field that haven't even been evaluated for toxicity in any meaningful manner—and are not likely to be so investigated. Need we point out the uphill battle to introduce rationality into the pesticide-agriculture scene, including the questionable antics of the Agriculture Department and State Legislatures throughout the country?

Need we point out the charade of the National Academy of Sciences appointing primarily atomic energy-supported scientists to re-investigate the hazards of ionizing radiation — men who have publicly taken a position on the matter at the outset of their supposed "study"? Suppose they do come out with recommendations suggesting a slight tightening of radiation standards. Is this a significant step forward in avoiding atomic energy depredation of the environment and of human heredity?

The creation of Centers for Adversary Assessment of Technology can fill an important void—can perhaps provide the "other side of the picture" of the hazards and secondary effects of technology at an early phase, before too much economic and ego commitment has occurred for a particular enterprise. Such adversary assessment is an absolute "must" for on-going and proposed technologies. It would be required for any proposed solution, since the "other side of the picture" is an absolute necessity. But unless additional steps are taken, the information developed by the adversaries will be arrayed against very powerful vested interests in all of the areas we've discussed. There is an additional element needed, ultimately, for the adversary activity to function effectively. And that element is conversion, in its broadest sense.

Conversion

Industrial conversion from manufacture of war materiel is receiving serious consideration. Obviously, it is highly desirable to encourage industry to cooperate in devising procedures that will make it acceptable *not* to push and lobby for unnecessary, destructive military expenditures. But this is far, far from enough. We must view conversion much more broadly and be prepared to encompass all types of industrial-technological endeavor — wherever it becomes evident that anti-societal goals are being pursued, no matter how innocently.

The fundamental premise has to be that industrial-technological endeavors directed toward improvement of the quality of life are necessarily preferable to those which contribute to ecocide. And a second premise is that we must absolutely learn to accomplish transition of anti-societal to pro-societal endeavor soon.

* * * * *

Indemnification: At the economic-entrepreneurial level, the necessary ingredient is *indemnification* against loss of investment when technology assessment dictates a change in direction. We would hardly be impressed by those economists who would say this is unrealistic, impractical, and unworkable. These same economists have failed in the past to include the secondary, and severe, costs to health and environment in their balance sheet thinking about corporate economics. If our suggestions *remain* unworkable or impractical, it will be because the economists fail to accept a major challenge which faces them to work out details that will be workable. The ultimate in economic stupidity is the degradation and destruction of life.

In at least two major areas the industrial entrepreneurs arrived at the position they are now in through public and governmental urging. We are not unmindful of complicity by the entrepreneurial lobbies in creating the governmental "urging." Nevertheless it is clear that the public and government did support the cold war concept and did, thereby, help create the vast military industry. Another illustration is in the field of atomic energy. There is little doubt that the Congress and the

Atomic Energy Commission worked hard to "sell" industry on the wonders of the peaceful atom, especially the wonder of nuclear electric power production.

Why would it not be proper to indemnify industry investors against capital loss required by a change in direction? Indeed, a failure to do this may well make it harder in the future to get industry to participate in governmental sponsored areas, some of which, at least, may be quite worthwhile. A punitive approach to investors in technologies which prove to be unwise can only be expected to meet with fierce resistance, subterfuge, distortions, half-truths, and lies in the effort to preserve parochial, short-term economic interest, whatever the societal cost. Far better to meet this problem by learning some economics of indemnification.

*   *   *   *   *

Preservation of Technologists' Positions: It is equally obvious that we cannot afford the luxury of unemployment or prospective unemployment for technologists, or for the labor force which is involved in their technology. For the first group, the technologists (and scientists), the prospect of the disappearance of their technology, their careers, their positions is, perforce, terrifying. Therefore, objectivity in their own assessment of their particular technology is readily buried in a morass of rationalizations and pseudo-science. The second group, the labor force involved, provides an unfortunate lobby to prevent public objective evaluation of the technology and its hazards.

We must develop techniques to protect both groups against unemployment and the fear of unemployment, if we are to expect them to participate in a constructive redirection of technology where required. Some economists have a tunnel-visioned view of unemployment as a useful tool in curbing inflation. Anachronistic and inhumane though this be, the implication is far, far more serious in a technology-based society. Obviously, where position and total career loss threatens, the technologists and the backup labor force will opt, overtly and covertly, for continuation of an anti-societal enterprise. And they will represent a powerful force to preserve the enterprise by delaying and confusing the hazard issues. Why should we stimulate this behavior—a behavior so human and expected?

We propose, therefore, when a technological enterprise needs cessation or redirection, that the technologists and labor force be guaranteed continued employment in the redirection of their particular technology. Again, the classical economist may argue that the expense would be prohibitive. And our answer is that failure to guarantee against position and economic loss will be *infinitely* more costly for society.

Certainly the legal profession has learned very well the difficulty of getting expert witnesses from within technology to testify concerning hazards of their technology. And they hope that somehow this wall of silence can be broken so as to be able to carry forward environmental lawsuits. Such hopes are, broadly, destined to failure unless the fear motive is removed. And that fear rests in economic and position losses, or potential losses.

Moreover, it is manifestly ridiculous even to consider unemployment for technologists and scientists (actually for anyone, for that matter). There are indeed many important tasks requiring all of our technological skills and ingenuity. Why waste it? There is little doubt that most technologists can readily be redirected into new areas. The cost of those who perform poorly during the redirection phase would be a small price to pay for the tremendous gains achieved by stopping eco-mad endeavors. And, further, technologists, realizing that redirection would be expected in the course of their careers, would be far less likely to become overly limited in confining their expertise to minutia of a specific technology.

*   *   *   *   *

Ego-Prestige Loss and Defensiveness: We all are familiar with the expression that "nothing succeeds like success." It seems like a homey little statement, until one considers carefully some of the implications. And this leads us directly to consider some extremely important issues *other* than the economic ones in the persistence of technological blunders.

We must ask ourselves seriously about the price of *failure,* rather than success. As a culture, we place a high ego-premium on being right about what we say, what we do, for essentially all endeavors that are in the public or semi-public domain. It is no secret that in scientific academe some men appear to devote a life-time of research and publication to proving they were right in their Ph.D. thesis. Who in industry or technology is unaware of the hazard inherent in having to tell his superior that all is not so rosy in the picture painted last month or last year concerning a specific project?

Defensiveness is the obvious result of the high value-premium we place upon success. And defensiveness breeds tunnel-vision, self-deception, and rationalization—anything *but* objectivity. Why can't we learn to honor and respect honest admission of error, of failure? While this may require careful nurturing of a subtlety in attitudes, we will fail to learn to change our attitudes at great peril and cost.

Decisions to go forward in a technological enterprise are not made by bureaus, nor are they made by corporations. Decisions are made by men. It is, of course, entirely appropriate to emphasize this in our endeavor to impress upon men that they will be held accountable for their decisions. This will certainly help in making captains of industry, governmental decision-makers, and technologists exercise more sobriety than otherwise they might. But at the same time we must absolutely refrain from squeezing men into an escape-proof, irrational ego-box.

Responsibility, yes — but only if we add sincere appreciation and praise for the ability of a man to admit forthrightly that he has changed his view, that what once looked right, now looks foolish. We desperately need to create an atmosphere where a man can proudly admit error. In a vast majority of instances the error is not the result of negligence, not the result of deceit, not the result of irresponsibility. It is simply the result of the great power of hindsight, especially hindsight buttressed by new evidence and altered circumstances. So, we had better see to it that something else can succeed besides success.

Time Magazine (December 28, 1970) carried a short article entitled "Heresy in Power." The title itself is extremely revealing of our attitudes. Presented in the article is the statement by Charles Luce, Chairman of Consolidated Edison Corporation, that the idea of promoting increased electric power use, representing the wisdom of three years ago, is the "idiocy of today." Why is that revised view of Mr. Luce regarded as "heresy?" It is, rather, a profoundly important realization by a leader of industry that his industry's position of yesteryear is no longer compatible with the real world. And, therefore, his statement deserves praise and admiration. Are we broadly prepared to provide such praise? Mr. Luce seems inordinately capable of learning and forthrightly stating the new horizons opened up by his self-education.

Thus, instead of the totally defensive attitude of the electric utility advertisers and AEC officials who whitewash the hazards of nuclear power generation because of their commitment thereto, Mr. Luce suggests the highly constructive idea of a tax on electricity bills to provide funds for research and development of new methods of electric power generation compatible with the environment. Since Mr. Luce is thoroughly familiar with nuclear power (his company participates in nuclear power), we can surmise that he refuses to be brainwashed concerning the absolute wonders of that approach to power generation. How many men can escape the irrational ego-box as well as Mr. Luce has? What reception will he receive for his "heresy" among his electric power colleagues?

APPENDIX I

NUCLEAR POWER
QUESTIONS AND ANSWERS

For Citizen Action Groups

The proponents of nuclear power conduct a well-financed campaign (largely paid for by us taxpayers) which relies heavily on a barrage of statements that minimize the hazard of nuclear electricity generation and extol its wonders. Citizen action groups need accurate, honest information to rebut such statements. Because so many of the questions concerning the value and safety of nuclear power recur over and over again, we have summarized the "questions and answers" and offer them here for ready reference.

QUESTIONS and ANSWERS

Setting Permissible Doses of Radiation

1. **How much radiation do AEC standards allow Americans to receive as a result of peaceful uses of the atom?**

   Answer: The maximum allowable whole-body dosage at the perimeter of a nuclear installation is 500 millirads to any individual. The average dose allowable for the U.S. population is 170 millirads per year. In fact, part of the justification for the *average* allowable dose being 170 millirads per year is that it was considered that this limit would prevent any *individual* from receiving over 500 millirads per year. (Millirads are equivalent to millirems.)

2. **Is the allowable or permissible dose of radiation truly safe?**

   Answer: Numerous atomic energy proponents repeatedly

make the claim that the "permissible" dose is a safe dose and they imply that no one will be injured at this "permissible" dose. However, there is *not a shred* of evidence that supports this claim of safety.

3. **What would the effects be for exposure at the "permissible" dose?**

Answer: The two major effects of concern are an increased death rate from cancer plus leukemia and an increase in genetic mutations.

Cancer + leukemia: For an individual steadily receiving 500 millirads per year, the chance of dying from cancer or leukemia is increased by 30 percent.

For a population averaging 170 millirads per year steadily, there will be a 10 percent increase in death rate from cancer plus leukemia. If this average were reached for the entire USA, there ultimately would be 32,000 extra deaths from cancer plus leukemia annually.

Genetic Risk: The genetic mutation rate would be increased by 5 to 50 percent for a population averaging 170 millirads per year. Ultimately this would translate into a 5 to 50 percent increase in death rates due to genetically-determined diseases. At the time such genetic deaths will be occurring, our population is expected to be 300 million persons. This means we can then expect between 150,000 and 1,500,000 extra deaths *each year.*

4. **How can a dose which leads to such high death rates be considered "permissible"?**

Answer: There is no justification whatever for such "permissible" doses, which can lead to public health disaster. So far as can be ascertained, the justification is simply that the development of atomic energy is not too seriously inconvenienced by such allowable doses. No health justification would appear possible.

5. **Are these "permissible" doses already being received by the U.S. population as a result of nuclear energy developments?**

Answer: Definitely not. It is fortunate that we recognize the serious health hazard of such allowable doses. We have, therefore, the opportunity to reconsider the wisdom of rash proliferation of nuclear electricity plants and other nuclear energy programs before an irreversible public health calamity has occurred.

6. **Is there scientific controversy concerning the hazard associated with "permissible" doses of radiation?**

Answer: The real controversy is political, *not* scientific. If the sound public health principles agreed to by such bodies as the International Commission on Radiological Protection and the National Committee on Radiation Protection are applied, it is not possible to reach conclusions significantly different from those listed above (Question 3). Scientists do differ from one another, concerning the precise magnitude of the hazard, in a minor way. But all, applying sound public health principles, would agree the hazard is large.

The electric utility industry has been misled by the AEC and the JCAE into believing that a "permissible" dose of radiation is "safe."

7. **Is the hazard of radiation exposure now recognized as greater than it was thought to be when the AEC standards were put into force?**

Answer: The hazard of developing cancer is now recognized to be about *twenty* times greater than it was thought to be when the AEC standards were set. This is the result of a totally unsound public health approach which has characterized every aspect of standard-setting.

The genetic hazard is also now recognized to be far

274

greater than was thought to be the case when the standards were set. In this case new medical information concerning the genetic basis of many major diseases is the reason why the hazard is now recognized to be much more severe than previously thought.

8. **What is meant by unsound public health practices in setting radiation standards?**

Answer: Evidence based upon experimental animal studies, available for 25 years, would have led to the serious expectation of cancer plus leukemia that we now realize we are facing with the "permissible" radiation doses. In the case of atomic energy, we have failed utterly to apply sound public health principles.

Obviously we shouldn't have made the error of demanding the human corpse at all. If such grossly unsound public health practices were extended to all poisons in the environment, we would indeed face a sorry plight as a species.

9. **When the FRC, ICRP, or NCRP (standard-recommending bodies) sets a standard for public exposure, do they mean to suggest that such exposures are safe?**

Answer: Absolutely not. These various bodies have never stated, nor implied, that such exposure is *safe*. This implication is a misuse of the standards by AEC, JCAE, and the electric utility industry. The net effect of such misuse of standards is deception of the public.

What the standard-recommending bodies *hoped* was that the benefits of the atomic technology might offset the cancer, leukemia and genetic hazards.

10. **Can it be demonstrated that such a body as the International Commission on Radiological Protection did not mean that anyone should construe their suggested limits to be safe?**

Answer: It most certainly can be demonstrated. The following is a direct quote from ICRP Publication 9 (1965).

"Because of the need for guidance in this regard, the Commission (ICRP) in its 1958 Recommendations suggested a provisional limit of 5 rems (5,000 millirems) per generation for the genetic dose to the whole population, from all sources additional to natural background radiation and to medical exposures. The Commission believes that this level provides reasonable latitude for the expansion of atomic energy programs in the foreseeable future. It should be emphasized that the limit may not in fact represent a proper balance between possible harm and probable benefit, because of the uncertainty in assessing the risks and the benefits that would justify the exposure."

So, the ICRP admitted forthrightly it didn't know the risks.

And the ICRP admitted forthrightly it didn't know the benefits. In the face of both uncertainties, they (ICRP) went ahead to "provide reasonable latitude for the expansion of atomic energy programs." Most people, including the electric utility people, were led to believe the standards were "safe," when the ICRP in fact said no such thing.

Standards arrived at in this manner can in no way be regarded as *safe* with respect to damage either to humans of this or future generations. Instead, such standards mean a moral judgment has been made that a certain number of human lives lost is acceptable for the good of the atomic industry. The entire standard-setting procedure is probably illegal and unconstitutional involving, as it does, the moral judgment to sacrifice the lives of humans explicitly.

11. **Is the entire concept of "safe" or "permissible" radiation exposures subject to challenge?**

Answer: Most certainly. Standards, as they have been set up to the present time, have nothing to do with the protection of the public health. They represent a set of numbers drawn out of thin air, for the convenience of a particular technology. The public and (in the case of atomic

276

energy) the electric utility industry are misled into believing that the words "standards" or "allowable" or "permissible" mean "safe." This is simply false.

12. **Would it ever be possible to set truly safe standards for radiation exposure?**

Answer: There is only one set of circumstances under which truly *safe* standards could be set. That would be if we *knew* that *some* amount of radiation was free of such harmful effects as cancer, leukemia, or genetic damage.

It can be stated unequivocally, and without fear of contradiction, that *no amount* of radiation has ever been *proved* to be safe. As a result, it can be stated unequivocally that there does not exist *any* standard which implies a *safe* amount of radiation.

13. **Does this mean that all codified "standards" represent trading human lives for some supposed benefit of technology?**

Answer: Unfortunately, yes. Worse yet, there has never been an effort made to demonstrate that the supposed benefits accrue to *those who suffer* the hazard. Nor has society even been permitted to participate in such an important decision.

Natural Radiation

14. **Aren't we exposed to radiation naturally?**

Answer: Yes.

15. **Is there any harm from natural radiation?**

Answer: Every responsible body (ICRP, NCRP, FRC) has explicitly stated that since we cannot prove that *any* radiation is safe, we *must* figure that harm in the form of cancer, leukemia, and genetic deaths is occurring from *all* sources of ionizing radiation in *direct* proportion to the amount of radiation received.

Since this is the basic underlying assumption for all responsible scientists, all of them would necessarily estimate that natural radiation, like man-made radiation, will produce its proportionate share of cancer, leukemias, and genetic deaths.

16. **There are areas in the world (Kerala, India and Brazil) where "natural" radiation is many times higher than in most other regions. What do these areas show with respect to cancer or leukemia induction by such natural sources of radiation?**

Answer: No adequate studies have ever been carried through in these regions to learn about the cancer and leukemia production by the high natural radiation. Such studies would be difficult, for they require careful control observations and large numbers of subjects.

For some strange reason the absence of studies is commonly equated, by atomic energy promoters, with an absence of harmful effect of the radiation.

17. **What did we do in assessing the cancer plus leukemia hazard for atomic energy?**

Answer: The standard-setting bodies refused to accept the experimental animal evidence, a grave public health blunder on their part. Instead, they demanded seeing *human* cancer and leukemia before they would consider human cancers and leukemia to be caused by radiation.

18. **Do we have the human evidence now concerning cancer?**

Answer: Unfortunately, yes. When the standards were set, the humans exposed to radiation (the Japanese A-Bomb survivors and 14,000 medically irradiated British subjects) had not been observed long enough for the radiation-produced cancers to develop. The result, the cancer hazard of radiation was estimated approximately twenty times too low. The new evidence showing the twenty-fold higher hazard became available just through the passage of enough time.

19. **It has been stated that even if the cancer and leukemia production by radiation is as serious as has been estimated, the average person will suffer only a shortening of life measured in weeks or months. Should we worry about such a "minor" effect on life expectancy?**

Answer: The correct answer to this question is to point out that this approach, considering average life expectancy, is truly immoral. What should really be considered is the shortening in life expectancy for those who suffer the cancer or leukemia from the radiation they receive. Many of these individuals lose 5, 10, 15, 20, or 25 years from *their* life expectancy as a result of the early death from cancer or leukemia. It is small comfort to these individuals or their families that their loss of life expectancy is made falsely to look insignificant by being averaged in with the life expectancy for those fortunate enough to escape the radiation-induced cancers or leukemias.

This can be illustrated by consideration of a specific situation. If a group of ten-year olds were to be irradiated, we know that a certain number of them will die of cancer or leukemia, the number who die increasing with each increase in radiation dosage. For those who *do* die of radiation-induced leukemia or cancer, the deaths start, after a latency period, some five years after the radiation, and new deaths are added annually for many years thereafter. Since the group of 10-year olds are representative of the population-at-large, their life expectancy *without* radiation should be some 50 to 60 years. But for those whose leukemias and cancers develop between the 5th and 10th year after radiation exposure, there has been a loss of 40 to 50 years of life, a matter of grave seriousness to them and their families.

It is difficult to understand the logic of those who treat

this problem by averaging the radiation victims in with those who escape the effect. By their logic the crime of murder is not a serious matter in the civilian population. After all, if we average in all those who are not murdered, the "average" loss in life expectancy for society is trivial.

20. **How much harm is natural radiation causing?**

Answer: Since at sea level we get approximately 100 milli-rads per year from natural radiation, we can calculate the harm very simply by direct proportion.
For cancer + leukemia

We estimated (see text) that exposure to 170 millirads would cause 32,000 extra cancer plus leukemia deaths per year.

Therefore, natural radiation *is* right now causing $\frac{100}{170}$ x 32,000 = about 18,800 cancer plus leukemia deaths each year. (Actually, since some of our population lives at a higher altitude and hence gets more than 100 millirads per year, the true number of cancer plus leukemia deaths must be higher than 18,800.)
For genetic injury

We estimated (see text) that for 200-million people, 170 millirads would lead to between 100,000 and 1,000,000 genetically caused deaths per year. Therefore, for natural radiation, we calculate $\frac{100}{170}$ x (100,000 to 1,000,000) = 58,800 to 588,000 extra genetic deaths *are* being experienced each year as a result of natural radiation.

21. **Is it really correct to refer to such deaths as "are being caused"? Have these deaths actually been observed?**

Answer: Adherence to responsible public health principles leads one to make the statement that these death *are* occurring. These are the numbers arrived at by applying the *ground-rule assumptions* that all responsible scientists and radiation study groups agree to follow, for public health purposes. If one denies these calculations, one is directly and overtly denying the sound principles of public health.

These deaths have not been observed as the specific ones caused by radiation. The radiation-induced cancer cases look just like other cancer cases. It is not relevant to ask whether the cases have been observed. It would be *folly* to consider such deaths as not occurring.

22. **Would it be possible to set up scientific studies to observe deaths from natural radiation by direct observation?**

Answer: It would represent a monumental scientific study to make the direct observations, but it *could* be accomplished with herculean effort. Unfortunately, technology promoters, especially atomic energy promoters, seem falsely to equate' 'no adequate studies done" with "no effect occurring."

The Linear Theory—
Damage Is Directly Proportional To The Dose

23. **What is the linear theory of relationship of radiation dose to effects, such as cancer, leukemia, and genetic injury?**

Answer: All responsible bodies involved in considering radiation hazards *have agreed* to use the linear theory to estimate the number of deaths caused for each amount of radiation. The linear theory holds that if *one* unit of radiation produces one case of cancer, two units of radiation will produce two cases, ten units will produce 10 cases of cancer, etc. In other words, the linear theory states that the damage is directly proportional to the dose *right down to the lowest doses*. Thus, it is obvious that if the linear theory is used, there simply cannot exist such a thing as a "safe" dose of radiation.

24. **But do all scientists agree with the linear theory?**

Answer: There is virtually nothing in the world upon which *all* scientists agree. But the question is not relevant. What is relevant is that every responsible scientific body and every responsible scientist involved in public health considerations of radiation injury *have all agreed* to use the linear theory for estimating cancer, leukemia, and genetic deaths until and unless someone *proves* otherwise. If someone claims to abide by responsible public health ground rules, *which include* using linear theory, and then says that some dose of radiation is safe, he is guilty of public health irresponsibility.

25. **Is there direct experimental evidence on animals or humans to support the linear theory of radiation injury?**

Answer: There most certainly is abundant evidence both for experimental animals and for humans—providing extensive support for linear theory in the production of leukemia, cancer, and genetic injury. And as new evidence is published, the experimental support for the linear theory becomes overwhelmingly strong. Listed below are major new reports supporting the linear theory with extensive scientific evidence for cancer and leukemia production by radiation.

(1) *Production of breast tumors in rats by x-rays and gamma rays.* (C. J. Shellabarger, V. P. Bond, E. P. Cronkite, G. E. Aponte, p. 161 in "Radiation-Induced Cancer," A Symposium of the International Atomic Energy Agency, Vienna, 1969.)

(2) *Production of bone cancer in mice by radium.* (M. P. Finkel, B. O. Biskis, P. B. Jinkins, p. 369, ibid)

(3) *Production of lymph cancer in mice by gamma rays.* (A. C. Upton and coworkers, p. 425, ibid)

(4) *Production of cancer and leukemia in children irradiated by x-rays while in utero.*
(A Stewart and G. W. Kneale, *Lancet,* p. 1185, June 6, 1970)

282

(5) *Production of Leukemia in Japanese Survivors of the Atomic Bombing of Hiroshima and Nagasaki.*
(S. Jablon and J. L. Belsky—Presented at the Tenth International Cancer Congress, Houston, Texas, May, 1970)

(6) *Production of Thyroid Tumors in Children irradiated with x-rays.*
(L. H. Hempelmann, *Science 188,* 160:159-163, 1968)

(7) *In the field of genetic injury, even the most optimistic data for genetic mutations, for slow delivery of radiation, shows linearity between dose and effect.*
(W. L. Russell, *Nucleonics 23,* No. 1, 53-62, 1965)

26. **Does the AEC agree to abide by sound public health principles?**

Answer: The AEC presents a remarkable paradox, caused undoubtedly by its dual role as promoter of nuclear energy and as its own regulator. The AEC claims to accept the guidance of the ICRP, NCRP, and FRC concerning radiation injury. All of these bodies use the linear theory as a fundamental ground rule of sound public health evaluation of radiation hazard.

The AEC is constantly adhering to two hopelessly inconsistent statements. For purposes of suggesting its public health responsibility, AEC accepts these ground rules. At one and the same time AEC suggests that 170 millirads of exposure is "safe." These two statements are absolutely irreconcilable. The reason why the AEC finds itself in such a difficult position is that if it really accepts sound public health principles, it will be led to estimating tens of thousands of extra cancers and leukemias per year for radiation doses which it considers "permissible." The public impact of this is horrible for the AEC to contemplate, so it is locked into an impossible and hopeless quandary.

The only escape for the AEC would be to admit the

truth to the public—namely, that its "permissible" radiation dose is really not at all safe. It is conceivable that the public may wish to accept a large number of extra cancer and leukemia deaths in exchange for atomic energy programs. Sooner or later the AEC will be forced to stop hiding this massive inconsistency in its position.

27. **But isn't it true that atomic energy programs, such as nuclear electricity generation, will deliver radiation slowly and that this might offer some protection?**

Answer: No acceptable evidence exists that slow delivery of the radiation will afford any protection against cancer or leukemia. Therefore, all responsible bodies (ICRP, NCRP, FRC) agreed to use the sound public health principle that *no* protection will be assumed for slow delivery of the radiation until and unless such protection is proved beyond doubt. Hence, *only* the total dose of radiation can be considered to matter. For anyone to claim atomic energy program "allowable" doses to be safe because of *slow* delivery of radiation is to violate the agreed upon fundamental public health ground rules. And this represents public health irresponsibility.

In addition, no acceptable experimental animal data indicates that cancer or leukemia will be lessened if radiation is delivered slowly. What experiments have been done simply show that radiation spread over a long period of time *may,* in some cases, produce less cancer than when the whole dose is given early in life. All this really proves is that young animals are more susceptible to radiation-induced cancer than old animals. Young humans are more sensitive than older humans, too.

28. **What is the net effect of the combined statements concerning linear dose versus response and the absence of proved protection from slow delivery of radiation—with respect to cancer and leukemia hazard?**

Answer: As a result of these public health ground rules,

all responsible bodies say that you must expect injury in direct proportion to the dose of radiation received. They all agree that, for public health prediction and action purposes with respect to atomic energy development, *no amount of radiation is safe.*

29. **Is it not true that the Russell mouse genetic studies show good evidence that slow delivery of radiation produces one-third as many mutations as fast delivery of radiation?**

Answer: At very high total doses of radiation the Russell studies do show one-third as many genetic mutations for slow delivery of radiation compared with fast delivery. For low total doses there is likely to be very little difference between slow and fast delivery of radiation. Since atomic energy programs will generally involve *slow* delivery of radiation, it is appropriate to explore the genetic consequences of radiation for slow delivery of radiation. This is the *most* optimistic possibility.

If we use the *most* optimistic Russell mouse genetic data, and even if we give full credit for slow delivery of radiation, we reach the conclusion that 100,000 *extra* genetic deaths per year would occur for the *allowable* average exposure of 170 millirads to the population. This can hardly be construed as an "optimistic" outlook, or a "safe" dose of radiation.

30. **Is it valid to transfer the Russell mouse genetic results directly to man?**

Answer: Of course not. Even the most optimistic outlook, based upon direct transfer of the most favorable mouse results, leads to the gloomy outlook for 100,000 extra genetic deaths per year for "allowable" radiation doses.

However, Russell himself proved that the mouse is *fifteen* times as sensitive to radiation-induced mutations as is the fruit fly (Drosophila). Man may, in turn, be much more sensitive than mouse. We simply don't know. If man *is* more sensitive than mouse, then the anticipated 100,000 extra genetic deaths per year will *rise* in proportion. How many times 100,000? No one knows.

Furthermore, Russell's own data show that some mouse genes are easier to mutate than others through ionizing radiation. If the critical genes in the human should turn out to be more readily mutated by radiation, then the 100,000 extra genetic deaths per year could rise appreciably.

Disparity of Estimates on Radiation Risks

31. **Is it true that the risk of radiation induction of cancer plus leukemia was seriously underestimated by standard-recommending bodies such as ICRP?**

Answer: Unfortunately, it is true. By refusing to use the long available experimental animal data, bodies such as ICRP committed a cardinal error that led to a *gross underestimate* of the cancer hazard from radiation. The ICRP used human data (from Japan and Britain) for the *early* period after irradiation, *before* the bulk of the radiation cancers had arisen. As a result, in ICRP Publications 8 and 9 (Pergamon Press) the ICRP estimated *one* extra cancer for each leukemia produced by radiation. We now know the *true* number is much closer to *twenty* extra cancers for each leukemia caused by radiation.

32. **Has the ICRP recognized its error in estimation of the cancer hazard from radiation?**

Answer: It certainly has. In a more recent (1969) publication, ICRP 14, the correct numbers are provided by a Task Force of the International Commission. It is extremely simple to show how these newly published numbers give such a greatly increased hazard of cancer from radiation.

By actually counting extra cancer and leukemia deaths in the 14,000 British subjects treated for arthritis by x-rays, Court-Brown and Doll (quoted in ICRP 14) have shown 5.3 extra cancers for each leukemia produced by radiation.

Next, the ICRP Task Force pointed out two corrections that are absolutely essential to arrive at the real cancer hazard from radiation.

(1) Only 40 percent of the bone marrow was irradiated in these 14,000 British subjects. (The leukemias arise primarily in the bone marrow.)

(2) The organs developing the extra cancers from radiation received approximately 7 percent as high a dose as the spinal marrow. Obviously, to compare cancer and leukemia one must correct the data so they are based upon the *same* radiation dose. The arithmetic is exceedingly simple.

First, we must correct the leukemia estimate for the fact that only 40 percent of the marrow was irradiated.

Thus, to correct to total body irradiation, there would be $\frac{100}{40}$ x 1 = 2.5 leukemias for every 1 case observed.

Second, we must correct the cancer estimate for the fact that the dose to the various organs developing extra cancer was 7 percent of that received by the spinal bone marrow. So, 5.3 x $\frac{100}{7}$ = 76 extra cancers for 5.3 observed, after correcting to the same total body radiation dose. Finally, to get the true ratio of radiation-induced cancers to radiation-induced leukemias, we simply take $\frac{76}{2.5}$ = 30. So these data, properly corrected by the factors presented in ICRP 14 lead to 30 extra cancers for *each* leukemia produced by radiation. This means that the earlier estimate (ICRP 8 and 9) had underestimated the cancer hazard by 30-fold just a few years ago.

Some observers feel that the factor of 7 percent for

dose correction (presented by Dolphin and Eve and quoted in ICRP 14) may result in an over-correction. Even if the extremely *rash* assumption is made that Dolphin and Eve were off by 100 percent, that is, the dose correction should be 14 percent instead of 7 percent, we would be led to estimate that extra cancers produced by radiation are *fifteen* times as frequent as is leukemia.

Incidentally, the ICRP 14 analysis doesn't take into account the fact that some organs weren't appreciably irradiated in the 14,000 British subjects. Therefore, for true whole body radiation, such organs will also develop cancers. Therefore, the 15 to 30 cancers for each leukemia is a minimum estimate of the hazard of cancer from radiation.

33. **The Gofman-Tamplin estimate of genetic deaths from exposure to "allowable" doses of radiation is 150,000 to 1,500,000 extra deaths per year for a population of 300-million people. How could the standard-recommending bodies have possibly chosen 170 millirads as an average population allowable dose to be acceptable in view of such estimates of genetic hazard?**

Answer: The erroneous thinking of the standard-recommending bodies was even worse on the genetic question than it was on the cancer plus leukemia question. Over a decade ago, these men *thought* they knew the kinds of genetic injury that would cause deaths. They were focusing on such uncommon single-gene diseases as hemophilia, galactosemia, phenylketonuria, and other rare diseases. Altogether these *rare* single-gene diseases add up to about one percent of all causes of death.

Since the standards were set, it has been discovered that *most* of the major killing diseases of humans have a genetic component, but it appears that more than one gene is involved. Such diseases are, therefore, called *multi-gene* diseases. Coronary heart disease, which kills more than *twice* as many Americans per year as all forms of cancer put together, is one such multi-gene disease. Diabetes mellitus, rheumatoid arthritis, and schizophrenia are other examples of multi-gene diseases. As a result of these discoveries concerning multi-gene basis for major diseases of our society, the *real* genetic hazard problem extends to between 50 and 100 percent of all causes of death, contrasted with one percent that was considered genetic when the standards were set.

Thus, the radiation standards were set with an *underestimation* of the genetic hazard by 50 to 100 times as a result of this error alone. It must be acknowledged that *new* knowledge of the past decade has led us to realize how erroneous the estimate of genetic hazard was when the radiation standards were set. But this is precisely a major point to understand. At any point in time, our medical and biological knowledge is fragmentary. And this means that standard-recommending bodies should lean far over to the conservative side, if they are to do even a minimum job in the field of public health protection.

34. **AEC spokesmen say the evidence for cancer plus leukemia comes from high doses of radiation, whereas the standards for peaceful uses of atomic energy are for much lower total doses. Doesn't this alter the hazard estimates?**

Answer: Not one iota. It has been pointed out in earlier questions and answers that every responsible body (ICRP, NCRP, FRC) has stated clearly and *repeatedly* that the linear proportionality *must* be used for public health purposes. Therefore, whatever is observed for high doses (say, 100 rads) will be expected to occur *in proportion* at low doses (say, 5 rads). Thus, if 100 rads produces 200 cancers, it follows that 5 rads will produce $\frac{5}{100}$ x 200, or 10 cancers. That's what the ground rules say, ground rules *everyone* accepts for hazard estimates. There is simply no way to escape this. AEC cannot continue to claim it will behave responsibly and accept sound public health ground rules and then turn around and use gibberish concerning high dose and low dose in attempts to obscure the real hazard.

Moreover, the statements that all the cancer-leukemia evidence comes from high doses of radiation are simply false. There are experimental animal *and* human data for low total doses, very damning data indeed. For example, Dr. Alice Stewart's data show that about one rad, a *very* small dose, delivered to the unborn child in the first 13 weeks of pregnancy will double the potential number of cancers and leukemias during the first 10 years of childhood.

35. **Over the years since 1945 a large number of workers within and outside the AEC have been exposed to ionizing radiation. Can we get direct evidence for the effect of radiation in producing cancer and leukemia from the records kept on such workers?**

Answer: Unfortunately there has been inadequate record keeping concerning the fate of workers so exposed. Indeed, only recently has the AEC introduced a procedure to insure that radiation exposure records will be maintained for periods of time long enough to be useful in this regard.

36. **AEC spokesmen say the 170 millirad standard for average population exposure is the "improper" standard. They say that 500 millirad (or 500 millirem) at the perimeter of the nuclear reactor is the appropriate standard to use. Further, they state that if this standard is used, it is impossible for the population average dose to ever come near the 170 millirem figure as a result of nuclear electricity generation. What is the response to these claims by AEC?**

Answer: In a simple statement, the AEC claims are irrelevant and simply fail to address the hazard issue at all. The AEC has picked out the *least* important part of the radiation hazard problem and dwells upon it. In this way the AEC takes note of only the very tip of the iceberg. They have thereby neglected *all the important* sources of radiation associated with nuclear power generation.

The perimeter dose of a nuclear power reactor *operating perfectly* is so small a part of the problem that it is hardly worth discussing at all. Therefore, we must examine the *real* problems of nuclear electricity generation.

(1) All authorities recognize food chain accumulation of a variety of radionuclides to be a major problem associated with the release of radioactivity into the biosphere. Even if the releases at the perimeter of a reactor were at the AEC "permissible" value, radionuclides that can go through the forage to cow to milk to human pathway can result in enormous multiplication of radiation dose in humans. Thus, contaminated milk can be consumed hundreds of miles away from a nuclear reactor with the result that people drinking such milk will get far higher doses than one would get by breathing contaminated air right at the reactor fence.

(2) The fresh water-to-fish pathway can concentrate radioactivity easily 1000-fold or more. Cesium-137 is an illustration of this. Thus, even though a water effluent at the release point may make the water drinkable with delivery of 500 millirems, the fish grown in such water, 1000 times as radioactive, can not be eaten in any quantity without grossly exceeding "tolerance" levels.

The AEC standards do not give proper attention to these food chain pathways, although the AEC is beginning to correct this error in new licenses, but *not* adequately.

(3) The AEC claim concerning population exposures from nuclear electricity generation neglects *all* of the following important sources of exposure:

(a) *Accidental* releases at the reactor. *No one* knows the risk of such accidental release for any of the currently planned reactors, since all of them are experimental reactors, large in comparison with those for which operating experience exists.

291

(b) *Accidental* releases during transport of spent fuel rods from the reactors.

(c) Releases and accidental releases at the fuel reprocessing plants. The one commercial reprocessing plant (West Valley, New York) has a very unfavorable operating experience. (see text)

(d) Releases and environmental contamination from low level and intermediate level waste releases and waste burial in the environment.

(e) Releases and environmental contamination from storage, burial, or final other disposal of the astronomically high level wastes left after fuel reprocessing. The handling of such wastes is still in the research and development stage, and the economics of such handling remain conjectural.

(f) AEC Commissioners themselves are on record saying they cannot be sure that the fuel rods in the reactors will not be leakier than expected from design specifications. Therefore, assurances of releases even under routine operation of new, untried reactors are, at best, conjectural.

(g) Accidental releases through sabotage at any step in the entire fuel and waste cycles are not even discussed by the AEC.

Safety Factor in Reactors

37. Don't some AEC and utility officials claim the reactors "safe"?

Answer: They do claim this. No one alive has any reliable estimate of the risk of major accidents in nuclear electricity generation simply because there is *no* valid operating *experience* for the current generation of large power reactors. Claims based on the handful of small power reactors are simply indefensible. Most reactor experts readily admit this. (see text)

38. **If we don't have enough operating experience, what is the justification for placing the new, untried larger nuclear electricity reactors near major population centers?**

Answer: There is no justification. A serious moral question is posed here. All AEC and electric utility officials should be asked to answer this moral question.

39. **If the AEC is so confident that nuclear electricity generation can go forward with trivial exposure to the population, why doesn't AEC itself argue for lower permissible doses and end all the arguments?**

Answer: This intelligent question has been asked by countless people at symposia, lectures, and in articles. They have yet to receive an intelligent answer from AEC spokesmen. It is, of course, obvious that if the AEC really believed the exposures would be as low as they *hope* they will be, AEC would assuredly jump at the opportunity to eliminate public concern by lowering the permissible doses. The only rational conclusion the public can draw is that neither AEC nor the electric utility industry has any confidence in these optimistic predictions.

40. **The AEC spokesmen repeatedly assure that the releases of radioactivity will be one percent of the permissible AEC limits. Can the public count on this assurance?**

Answer: If anyone *really* knew what the releases are going to be from the new, large, experimental nuclear plants, the problem would be very different indeed. But *no one* knows what the releases are going to be, either initially or after various time periods of operation. The best evidence on this can be found in the recent Hearings of the Joint Committee on Atomic Energy (Environmental Effects of Electric Power Production, Part I). When Commissioners Ramey, Thompson and Johnson were asked why such a large "cushion" was required in the form of *high* permissible standards when the releases were going to be so *low,* all three men answered truthfully that they simply didn't know what the releases were going to be for the new plants, and, therefore, the high allowable doses were required as a "cushion".

Obviously, anyone predicting radioactivity releases for new, experimental nuclear power plants is engaging in sheer speculation. It is unfortunate that the public is serving as guinea pigs in this gigantic, speculative experiment.

41. Are there additional reasons for concern over the burgeoning nuclear electricity industry?

Answer: The speculative character of the estimates concerning routine releases is only *one* reason for reservations concerning the construction of such plants. There are more compelling reasons.

(1) The entire nuclear electricity industry, including the electric utility industry, had been misled into thinking that a dose of 170 millirads had a wide margin of safety built in. We now know there was not only *no* margin of safety, but we know further that the cancer risk is some 20 times larger than thought when standards were set, and the genetic risk is some 50 to 100 times larger than thought when the standards were set.

Any time that an engineering development goes forward with the *false* illusion that a 100 or 1000-fold margin of safety exists compared with the real hazard, that entire engineering development can not be trusted in any way.

The appropriate first step is to stop construction of any further nuclear power plants. Then it is essential to educate the nuclear manufacturing industry and the electric utility industry concerning the *true* magnitude of radiation hazards. Once they realize the true magnitude of the hazard and they rethink their engineering in terms of the *realities* they face, reasonable discussions concerning nuclear plants can be held. Not before.

(2) On January 28, 1970, then Secretary of HEW, Robert Finch, ordered a complete review of all aspects of radiation standards. This review is only now getting underway, and it is estimated that it will take two years. The review is being conducted largely by atomic energy proponents, so that there is no doubt that the conclusions will be subjected to severe scrutiny by the ever-larger informed scientific and lay community. The radiation standards controversy may last well beyond the two years of the formal review, since there may be severe challenges to a review conducted primarily by atomic energy supported scientists.

No one knows, out of all this, what the new standards for allowable radiation exposure will be. Thus, if new nuclear power plants continue to be constructed under the obviously unacceptable present standards, it may be that such plants will be unable to meet the new standards. This may represent a colossal economic blunder for the electric utility industry, a blunder they would undoubtedly pass on to the electricity consumer in the form of rate increases. It would seem far wiser to have a moratorium on new nuclear plant construction until all aspects of the radiation hazards controversy are settled.

(3) In spite of numerous irresponsible assurances to the contrary, *no one* knows the risk of a catastrophic accident for the large, experimental nuclear power plants now being planned for and built near major population centers. There simply is *no* relevant experience upon which to draw. Dr. Walter Jordan, a nuclear expert, and a member of the AEC Atomic Safety and Licensing Board, has spelled this out in *no uncertain terms* in his article, entitled 'Nuclear Power, Benefits versus Risks', which is published in "Physics Today", May 1970. In that article Dr. Jordan admits we have no idea of the odds of a major accident because we have no experience. It is not reasonable to have the population of American cities become the guinea pigs for such experience.

Price-Anderson Act—
Insurance Against Personal and Property
Damage from Nuclear Accident

42. **Is it true that the American public does not stand to recover from damages suffered as a result of nuclear power accidents and damage to property?**

    Answer: Sadly, but unfortunately, this is true. Historically, we know that neither the electric utility industry nor the private insurance industry was willing to gamble money on the liability that might be incurred in a major nuclear power accident. An AEC report (WASH-740) itself indicated that for the earlier, much smaller, nuclear power installations, an accident could result in 7-billion dollars in property losses. Because of this, the nuclear electricity industry was at a standstill. The promotional Joint Committee on Atomic Energy pushed through Congress a notorious piece of legislation, the Price-Anderson Act, which, in effect, *absolved* the electric utility industry of real liability for major accidents. This Act, at the same time, removed any hope for the citizen to recover more than a nominal part of damages suffered. Thus, the Price-Anderson Act decreed that, no matter how severe the damages, the maximum to be paid out for a single nuclear power disaster is 560-million dollars. Simple arithmetic shows that for an accident causing 7-billion dollars of damage the citizen, at most, could hope to recover seven cents on each dollar lost.

43. **Why should the electric utility industry be freed of liability for damage its nuclear plants cause, when other industries have to stand behind their activities fully?**

    Answer: The only answer possible is that this represents a shocking disfranchisement of American citizens, perpetrated by those who wish to promote atomic energy at any price to the public. Certainly the repeal of the Price-

Anderson Act is an early item of the highest priority for citizens concerned about their lives and property.

44. **If the Price-Anderson Act is repealed, would the electric utility industry be willing to go ahead with nuclear electricity generation?**

   Answer: There have already been a number of statements by utility officials that they would not go ahead with nuclear power plants if they had to bear the financial liability for the consequences of major accidents. And we also know that the private insurance industry refuses to insure the full liability for major accidents. Thus, the ill-considered nuclear electricity industry would undoubtedliy come to a standstill if it had to be financially responsible, as all other industries must be.

45. **If nuclear electric power plants can not be insured for the full amount of damages they can produce in a major accident, why does the AEC, a governmental agency, allow such plants to be licensed at all?**

   Answer: If we had a rational society, it would be unthinkable for a governmental agency, such as AEC, to grant a license for building nuclear power plants that cannot be insured for the real damage they can wreak upon the public. But the Congress has acted very unwisely in giving a *promotional* agency, AEC, the power to license the wares it sells. Obviously there is an extreme conflict of interest here. The AEC, favoring its promotional role, goes ahead to license nuclear plants that are uninsurable.

46. **Can't the individuals in the public recover damages from nuclear power plant accidents through their personal homeowners' policies?**

   Answer: Not a chance. The private insurance industry, with a notable reputation for making money, realized the hazards of the advent of the nuclear electricity industry. They, therefore, moved swiftly to protect themselves, not the public. Thus, most homeowners' policies now have a nuclear exclusion clause absolving the insurance company of liability for home damage caused by nuclear accidents or radioactivity.

While the insurance companies are to be commended for their financial wisdom in protecting *themselves* against the nuclear electricity industry, it is sad that the public is almost totally unaware that this action has left them in a position to lose catastrophically. Every citizen should be informed about his loss of rights to recover for property damage.

47. **Is there any reasonable compromise on this issue of safety and insurance against nuclear power plant accidents?**

Answer: Yes, there is one excellent compromise. Stop licensing any nuclear power plants until the insurance industry can be convinced to provide *full* coverage for losses sustained from nuclear power plant accidents. Either the nuclear power plants are safe enough to be fully covered by insurance, or they are too unsafe to be licensed at all. There is no middle ground.

48. **It is said that the reason why private insurance companies refuse to cover the full liability of nuclear power plants is that they have no "actuarial" experience with such accidents upon which to base insurance rates. Is is this true?**

Answer: Absolutely correct. The private insurance companies are far too shrewd to accept the optimistic, but totally unsupported, reassurances concerning safety emanating out of the Atomic Energy Commission. The insurance industry, having no relevant experience, *refuses* to risk dollars. But the American public, also not having the relevant experience, are being *forced* to risk their lives in a gigantic experiment.

"Clean" Nuclear Plants vs. "Dirty" Fossil Fueled Plants

49. **Isn't it true that fossil-fueled plants also create a health hazard? Would it be wise to stop nuclear electricity plants and accept the hazard of poisonous emissions from the dirty, fossil-fueled plants?**

Answer: Dirty, fossil-fueled power plants are a national disgrace. Every citizen should be adamantly against being poisoned by them as he should be against being poisoned by deadly radioactive emissions.

The real truth is that this question is phrased poorly, representing a phrasing that is carefully sponsored by the public relations branch of the nuclear power industry. The nuclear power industry wants the public to think that the choice is between *dirty* fossil-fueled plants and nuclear power plants. This is simply ridiculous. The technology to stop the poisonous emissions from outmoded fossil-fueled plants is far developed, and could be installed in the near future, if the demand for this were insistent. It is up to the public to declare its outrage against such poisonous emissions.

We could have eliminated the poisonous emissions from old fashioned fossil-fueled plants long ago. But all the research and development funds that *should* have gone in this direction were siphoned off into atomic energy. This has been a grave error. There is no doubt that we can and should have *clean* power from fossil-fueled plants. We will have if we insist upon it. Public pressure is *the* effective tool to achieve this.

50. **Are there even more attractive alternatives than fossil fuel or nuclear power?**

Answer: There are two parts to the answer. The first and most important task is to introduce some reason in the dialogue concerning electric power requirements. The electric utility industry stampedes the public into nuclear power with the threat of brownouts and blackouts and at the same time spends an advertising fortune to *stimulate* the use of more electricity.

Mr. Charles F. Luce, Chairman of New York's Consolidated Edison Company, as recently quoted (Time Magazine p. 40, December 28, 1970) takes an eminently sensible position:

". . . But last week he told a startled Manhattan audience: 'The wisdom of three years ago is the idiocy of today.' Instead of trying to increase consumption (of electric power), he now wants to decrease it.

"Luce is regarded as one of the most socially responsible leaders in the utility business . . . (Conservationists) argue that power generation also generates pollution—and now Luce has publicly agreed with them. . . ."

Mr. Luce is certainly to be commended for raising "the serious question of whether we ought to be promoting any use of electricity."

The second part of the question of alternatives is equally important. Here again Mr. Luce has made a worthy suggestion: "As a long term solution, Luce last week suggested a new federal excise tax of 'perhaps 1%' on electric bills to speed new ways of generating power compatible with the environment." Here Mr. Luce has come to grips with the heart of the problem, which is to press forward with methods for electricity generation *compatible with the environment*.

Numerous attractive opportunities abound, including solar power, geothermal power, clean fossil-fueled power, increased *efficiency* of power plants, and fusion power. If some of our elegant scientific talent were stimulated to develop these *sensible* approaches, we would undoubtedly hasten an early ecologically-sound solution for electric power generation.

The first step is to break the stranglehold on energy research and development dollars that has been held by the super-promotional, narrow-visioned Joint Committee on Atomic Energy.

Scientific Objectivity and Radiation Damage

51. **The Chairman of the AEC, Dr. Glenn Seaborg, is a highly respected scientist. Can't we rely on him to take an objective view of radiation hazards?**

Answer: Dr. Seaborg is indeed a highly respected, capable scientist. Yet in recent speeches he has taken a position diametrically opposed to all sound public health principles in assessing radiation hazards. In effect, he has ridiculed sound public health practices in the following statement from his November 19, 1970 talk entitled "Power, People, and the Press." Let us quote Dr. Seaborg directly:

"Unfortunately, much of the alarm being generated today is based upon speculation that has its roots only in manipulating statistics — in large linear extrapolations. These make the assumption that if X number of people are killed by a certain high level of radiation, known numbers of other deaths can be deduced at each decreasing level. But as a prominent radiation expert recently pointed out, such extrapolations can become ridiculous, and especially upon reaching certain low levels. They are, in effect, saying that if 1000 people die in a 100-mile an hour hurricane, 100 people will die in a 10-mile an hour breeze.

"Though I admit this analogy is not a scientifically accurate comparison with radiation effects, such ridicule seems justified because there is much evidence of biological repair mechanisms that counteract the effects of low levels of radiation. . . ." (Glenn T. Seaborg)

Dr. Seaborg has, in this unfortunate speech, chosen to ridicule the sound public health principles adhered to by all responsible scientists concerned about radiation hazard and he has chosen to ridicule the sound principles adhered to by all responsible standard-recommending bodies, including ICRP, NCRP, and FRC.

The public should now address individual letters to

Chairman Seaborg and ask him to produce one iota of evidence that is convincing of *any* biological repair mechanism that will protect against cancer and leukemia from low levels of radiation. If he has such evidence, it has been thoroughly hidden from the scientific community.

52. **Why does Chairman Seaborg of AEC try so hard to ridicule the hazard of radiation?**

Answer: He does not like to admit that he made a mistake. None of us do. He and other atomic energy officials began an intense promotional campaign of promises about the wonders of nuclear electricity several years ago before the extremely serious hazard of radiation was appreciated. It is one of the most difficult problems any human must face to admit past errors and to face unpleasant realities.

Undoubtedly, the public would admire Dr. Seaborg's courage if he faced up to the issue squarely and said he realized now that the hazards of atomic energy programs are far greater than he previously thought. Everyone makes mistakes, and the public admires men who can forthrightly acknowledge mistakes. This is especially true when the mistakes were based upon erroneous knowledge of the past.

53. **Isn't the problem of defensiveness a serious problem of technology in general with respect to environmental problems?**

Answer: Absolutely. Until recently, technology was given a green light to develop hardware as rapidly as possible. Today we recognize that such technology can produce an unbearable environmental burden. Nuclear power may represent one of the best examples of this. How can we expect men whose professional lives, jobs, and fortunes depend upon a particular technology to view the abolition of their technology with equanimity? If they weren't defensive, they wouldn't be human. Job security and professional life work are very important to people.

54. **Is it being suggested that society foot the bill for ill-advised technological ventures?**

Answer: Representatives of our government pushed atomic energy upon industry. Many scientists and engineers were urged to prepare for careers in this field. Now with a greater awareness of our environmental crisis, all of us must share to some extent in the burden of a change in direction. Just as we must make accommodations (with tax money) for loss of jobs and careers in other fields, so must we assume some of the responsibility for tiding the Atomic Energy Commission over the setback that will occur when the building of nuclear power plants is postponed, until we know how to make them safe.

Why So Much Fuss Over Such a Little Radiation from Nuclear Plants?

55. **Nuclear power advocates commonly state that the radiation one might receive by living in the near vicinity of a typical operating plant site for an entire year is equivalent to about what one would receive on a single round trip, coast-to-coast airplane flight. If the radiation is so low, why is there any concern about nuclear power?**

Answer: Nuclear power advocates almost step over one another in their hurry to claim how *little* radiation one will get from nuclear power generation. This should be greeted with, "Bravo!" Since nuclear power advocates claim it is ridiculous to think any significant radiation will be delivered, the nuclear power advocates *should be the vanguard* in the fight to lower the allowable radiation doses to the public. So far, when such optimists are asked why they aren't fighting to lower the *allowable* doses, they are speechless. Just as soon as their *actions* begin to match their *words,* credibility might be assigned to them.

56. **Isn't it true that medical exposure to x-rays is providing more radiation dosage to the public than nuclear power at present (1970)?**

Answer: Yes. This points up two necessities:

(a) Every effort should be made to reduce the exposure in the course of medical diagnosis. This can be done and *is* being done. Dr. Karl Z. Morgan has courageously campaigned for reduction in x-ray exposure from medical and dental use of x-rays.

(b) We should be very happy to have discovered the serious hazard of radiation *before* the nuclear power industry has led to a prospect for irreversible pollution of the environment. It is very hopeful that we can stop the nuclear power stampede before it has done irreversible damage, rather than later.

57. **In a speech on May 21, 1970, the late Commissioner Theos Thompson stated the following:**

"Radiation is understood much better than almost any other of the possible effects caused by man or his environment. It is strange that we who believe that atomic energy is an improvement in our environmental situation find ourselves attacked on the environmental basis, when we know full well that when the final choice is made, nuclear power will prevail because the alternatives to nuclear power will have much worse effects on human health!"

What is the answer to such comments?

Answer: It is *because* we know now how hazardous radiation is that we should reject a stampede to nuclear electricity which is certain to increase radiation hazard. It makes no sense to plunge headlong into a technology where we *know* the hazard is large.

As to the hazard of alternatives, we most certainly should *not* accept dirty fossil-fueled power plants instead of nuclear plants. But it is only the limited vision of nuclear power advocates that results in their failure to appreciate that environmentally-sound alternatives *can* be and *must* be made available.

This is an excellent illustration of how a parochial interest leads to a limited vision concerning possibilities for alternatives. There are very few reasons for believing any *rational* society will make "final" choices in favor of nuclear fission power.

58. **In that same speech, the late Commissioner Thompson stated:**

"If a state agency arbitrarily lowers the levels which are permissible in the state until they are barely above normal levels, they run some risk that at some time they will either have to require shutdown of this plant, or else find some graceful way to back off from their own regulation."

What are the implications of this statement?

Answer: Commissioner Thompson was, in effect, saying that no one really knows how much radioactivity would be emitted by the new plants. Therefore, he was warning states against restricting releases of radioactivity.

Of course, his comments are correct. What he is saying is that the promises of the nuclear electricity industry concerning how little radioactivity they are going to release are simply empty, untrustworthy promises. In effect, it is fine for nuclear electricity advocates to *promise* low releases of radioactivity to the environment, but it would be unfortunate if regulations were enforced to make them adhere to their optimistic promises.

59. **In a speech (January 23, 1970) AEC Commissioner Larson stated the following:**

"During the 27-year period (of nuclear power development) there has not been a single nuclear reactor accident in which a member of the public has been injured in any way whatever. This safety record is unequalled in any other industry in this country. This experience covers the operation of 435 nuclear reactors."

What is the answer to this claim?

Answer: First of all the nuclear reactors of this period were almost all tiny compared with the giant plants now being built, so the "27 years" of experience is hardly meaningful. Second, no one really knows how much total radioactivity was released by all these plants, since very few of them were independently monitored. Third, and most important, the claimed safety is only *by* definition. The way the Atomic Energy Commission operates is to *deny* all the extra deaths due to radiation. Having denied the deaths, they then claim safety. How is it that they "deny" the deaths? Very simply. Radioactivity *has* been released from many of these nuclear power reactors. First, the AEC calls an amount of radiation that can produce cancer by the term "safe". Having done this, they deny culpability for any deaths caused in people who have received this *falsely* "safe" dose. So, many people can be killed by such "safe" doses of radiation and the AEC simply denies the deaths by saying, "How can these people have been hurt by radiation — they received no more than our 'safe' dose?"

Thus, suppose the U.S. population were to receive an average exposure of 170 millirads per year and were to build up to 32,000 extra cancer plus leukemia deaths each year due to the radiation. Commissioner Larson would, by his reasoning, still say nuclear radiation had caused no deaths, simply because he would *define* the 170 millirads as the "safe" dose, and he would therefore dismiss the extra 32,000 deaths each year as non-existent.

306

Is Nuclear Power the Only Answer
to Future Needs?

60. **AEC Chairman Seaborg stated (April 22, 1970) the following:**

> "It has long been recognized that nuclear energy's full promise for providing a virtually unlimited energy source for future generations could be realized only through the development and application of the breeder reactors."

Is this statement correct?

Answer: This statement of AEC Chairman Seaborg is a classic illustration of the parochial absence of vision that characterizes nuclear power advocates. Numerous responsible scientists have indicated that solar power, geothermal power, and fusion power have excellent promise to provide unlimited power with *minimal* environmental deterioration. The limited vision of AEC Commissioners prevents them from seeing these possibilities, as they continue to focus upon nuclear fission power.

One might add that with the plutonium hazard introduced along with the breeder reactor, it is problematical as to whether future generations would get to enjoy the "unlimited energy source."

61. **In that same speech, AEC Chairman Seaborg said:**

> ". . . it is the utility (industry) which has the basic responsibility to make sure that the plant is designed, constructed, and operated to meet requirements for safety, reliability, and economics."

Is it true that the electric utility industry carries these responsibilities, and that AEC is not responsible?

Answer: Sadly enough, this *is* true. And the consequences are horrible to contemplate. The electric utility industry

307

has been misled by the AEC into thinking the "allowable" doses of radiation have a wide margin of safety in them, when indeed we know there is *no* margin of safety. So the utilities have gone ahead with all their engineering with this false *illusion* of safety. While Dr. Seaborg is right that the utility industry will be the patsy for the deaths that will be produced, the real losers will be the unfortunate citizens who develop the radiation-induced cancers and leukemias.

Environmental Considerations with Nuclear Power Plants

62. **Chairman Seaborg continued:**

"But when we hear members of the scientific community express the view that despoilation of the environment might make life untenable on this planet by the year 2000, we can but wonder if they know something we don't know."

Is it really true that no one has told the AEC the true facts concerning radiation hazard?

Answer: It is regrettably true that members of the scientific community *do* know something the AEC does not know. The difficulty does not arise because the AEC hasn't been told about radiation hazards. The difficulty arises because the AEC refuses absolutely to listen to anything concerning radiation hazards.

63. **Another statement the AEC Chairman made in that speech:**

"In our efforts to bring about increased public understanding, I see one need as central and pervasive. That is the need to create a heightened awareness of the need for energy, of the extent to which it is essential to our welfare today, and of our increasing dependence on it in the future, particularly with respect to protecting the environment. . . ."

Is this really what is needed?

Answer: At a time when all those concerned with the environmental crisis have come to recognize the pollutional aspects of energy consumption and the wisdom of questioning increases in energy consumption, the AEC is 180 degrees out of phase and is, in effect, arguing in favor of *more* of the same approach that brought on the environmental crisis in the first place.

64. **In a speech (March 26, 1970) Commissioner Larson said:**

"Far from being one of the curses of the nuclear age, as many well-meaning laymen believe, the nuclear waste disposal problem is, in fact, one of the smaller such tasks in our modern high consumption society."

Is it really true that nuclear waste disposal is no problem?

Answer: Apparently so *for the AEC.* But a committee report of the National Academy of Sciences was severely critical of AEC waste disposal practices—practices which could lead to serious fouling of the environment. Indeed the report was so critical of AEC waste disposal practices that the AEC suppressed it for over three years (see text).

News accounts of AEC laxity in waste disposal practices appear repeatedly. The AEC is not yet convinced that the earth is finite in its capacity to serve as a sewer. Probably, this is why Commissioner Larson feels that the legacy of radioactive garbage is perfectly appropriate for us to visit upon future generations.

Informing the Public on Nuclear Power

65. In a speech (April 4, 1970) AEC Commissioner Ramey stated the following:

"Despite our efforts to inform and listen to the public, however, we have been charged with failure to inform the public or to consider the public's views in the setting, by rule, of radiation-emission standards for nuclear power plants. . . ."

What is the answer to Commissioner Ramey's statements?

Answer: There can be no doubt that the AEC has spent millions of dollars to "inform" the public. The only difficulty is that all the "information" represents a Madison Avenue huckster message, replete with fairytales about "clean, cheap, safe nuclear power." It is impossible to find any evidence that would indicate effort by the AEC to present to the public a *balanced* set of information concerning the radiation hazards question.

With respect to the public's point of view being heard, the record of AEC Atomic Safety and Licensing Board Hearings can only be regarded as a national disgrace. The *real* issue before such Boards should be:

(1) The validity of the primary radiation standards,

(2) The moral question of going ahead with nuclear power plants when *safety* cannot be assured. Neither of these questions has been admissible. The public has been limited to *proving* that the *design* specifications for the particular reactor do not meet the established standards. If it is the *established* standards that are really in question, it is the public's *right* to question these standards. The failure to allow this is the essence of the answer to Mr. Ramey's plaint. Indeed, the AEC deserves more severe criticism than it has received for its *manifest* failure to listen to the proper questions raised by the public, — a public acting appropriately to protect life and property.

66. **In that same speech, Commissioner Ramey went on to say:**

"Members of the public whose interests are affected can intervene in these hearings, and can call witnesses and cross-examine in order to try to satisfy themselves as to the safety of the proposed plant."

Is it true that the public has all these opportunities?

Answer: In principle, the public has these opportunities. In practice, the public is deprived of the opportunity in such a way as to mock our democratic society. First, it has been estimated that a successful intervention would require $100,000 to $200,000. Why should the public be forced to provide such funds to raise appropriate questions? Why should taxpayer dollars be utilized by the AEC to prevent the public from being adequately represented and heard? Second, the rules concerning *what* questions can be raised in such hearings make it impossible to raise *any* of the pertinent issues about which the public is *properly* concerned. These *really* pertinent issues are listed in the previous question.

67. **Later in the same address Commissioner Ramey stated:**

"Discharges of radioactive materials from nuclear power plants operating to date have generally been only a few percent of the AEC limits."

Is this true?

Answer: Commissioner Ramey's statement is an illustration of the one-sided, distorted picture presented by nuclear power advocates. It is true that many plants, under routine conditions, have operated at a few percent of AEC limits. But why does Commissioner Ramey not add to his statement that the Dresden Plant had a period of over a year at about *30* percent of AEC limits and the Humboldt Plant, for a long period, operated at between 55 and 60 percent of AEC limits?

Why does Commissioner Ramey not discuss the appalling state of affairs at the West Valley Fuel Reprocessing Plant, where *citizens groups* had to go out to measure and prove that the releases were far above "a few percent of AEC limits"? Why does Commissioner Ramey not mention that once the citizens groups discovered the sad state of affairs at this plant, their findings *were* confirmed by the Bureau of Radiological Health? It is the *omissions* by men like Commissioner Ramey that lead to the loss of credibility of the AEC.

Who Are the Critics of Nuclear Power?

68. Commissioner Ramey further stated:

"There are a lot of people going around the country spreading 'scare talk' about nuclear power. Some of these people whom I have called the stirrer-uppers, like Larry Bogart, apparently make their living by creating controversy without regard to fact. Others are scientists who haven't done their homework or who insist on broadcasting their theories without first subjecting them to the review of their peers."

Is there any validity to Commissioner Ramey's remarks?

Answer: None. Society has far more to fear from the proponents of nuclear power who resort to the type of slander evident in Mr. Ramey's remarks. Gofman and Tamplin offered to test the sincerity of those who *claim* they want a review of the evidence by peers. They did this on January 28, 1970 by offering to debate the radiation hazards question with the AEC before a jury of the most eminent scientific peers. This offer is in the Hearings of the Joint Committee on Atomic Energy *in print*. (Environmental Effects of Producing Electric Power, Part 1). The AEC has steadfastly refused to participate in consideration of the evidence *before a jury of eminent peers.*

Mr. Ramey must be fully aware of this offer. If he is *not* aware of it, then he, as a Commissioner, is the one "who hasn't done his homework" by failing to know what is in those hearings. Moreover, he was *present* at those hearings.

What can the public think of the AEC, having itself declined to accept a review before peers, subsequently criticizing others for failing to subject their findings to a similar review? Can it be that the AEC, and *not* its critics, fear the review of the evidence. The record is quite clear for all to examine.

69. **In a speech (February 9, 1970), Commissioner Ramey charged:**

"Gofman and Tamplin and others are . . . violating one of the cardinal principles of scientific endeavor by not subjecting their conclusions to the normal review of their scientific peers. Instead they are trying their cases in the press and other public forums. We used to call such characters 'opera stars'."

What is the answer to this charge?

Answer: It should be expected of public officials at the high level of AEC Commissioner not to voice such totally irresponsible statements. Unfortunately for the AEC, they destroy the credibility of the AEC, not the credibility of Gofman and Tamplin.

The record on this matter is clear. The Gofman-Tamplin findings were presented before a highly respected scientific forum, a Symposium on Nuclear Science before the Institute for Electrical and Electronic Engineering at San Francisco, October 29, 1969. It would be difficult to imagine a more appropriate forum. Let us quote from the published Transactions of that Symposium* (a highly respectable *Scientific* publication) why that presentation was *requested* by the distinguished Program Committee of that scientific society:

"Up until the last few decades, man's ability to pollute his environment was relatively slight. In recent years however, the combined population and technological "explosions" have magnified this hazard to awesome proportions. The engineering community has a major responsibility and role in understanding and controlling this situation before it is too late. The Plenary Session was most effective in presenting these problems through the presentation of five invited papers:

"Pollutants and Natural Minor Constituents of the Upper Atmosphere"—Dr. E. A. Martell, National Center for Atmospheric Research, Laboratory of Atmospheric Sciences, Denver, Colorado

"The Chemical Invasion of the Oceans by Man"—Dr. Edward D. Goldberg, Scripps Institution of Oceanography, La Jolla, California

"Stable Isotope Measurement of Sulfur Pollutants"—Mr. Bernard Manowitz, Brookhaven National Laboratory, Upton, Long Island, New York

"Low Dose Radiation, Chromosomes, and Cancer"—Dr. John W. Gofman, Associate Director, Biomedical Division, Lawrence Radiation Laboratory, University of California, Livermore, California

"Fast Breeder Reactors—Status and Prospects"—Mr. P. M. Murphy, Manager of Advance Engineering, Breeder Reactor Development Operation, General Electric Company, Sunnyvale, California

Unfortunately for those who did not attend the Symposium, it is only possible to include the last two of these excellent papers in this issue of the Transactions.

Gofman and Tamplin pleaded for the AEC to join them in a review before peers, and Commissioner Ramey knows this full well. He knows, too, that Gofman and Tamplin have published their findings in several scientific journals, with thorough review by "peers." Commissioner Ramey knows that his statement concerning "trying their case in the press and other public forums" is fatuous and as valid as a three-dollar bill.

---

*I.E.E.E. Transactions on Nuclear Science, February 1970, volume NS-17, No. 1. Part 1.

70. In that same speech Commissioner Ramey continued:

". . . we have made less (progress) with our non-scientific antagonists—the 'rag tag' stirrer-uppers such as Larry Bogart and others who use high school debating techniques which are sometimes surprisingly effective. They seem most vulnerable when asked what alternatives they might suggest to meet the increased needs for electric power. Some, like Bogart at the Senate appropriations hearings last fall, talk about some far-out alternatives such as MHD (magneto hydrodynamics) or harnessing the Gulf Stream, or using solar energy or even tidal power. One finds also a certain softness by them in relation to coal and fossil sources, and some obtuseness as to the air pollution problems of these competitive sources. I believe the ridiculous position of these fellows could be made more evident."

How shall we view these remarks of Commissioner Ramey?

Answer: Mr. Ramey speaks for himself with respect to his disdain for the lay public and the excellent questions they raise concerning alternatives for electric power production. Indeed, distinguished scientists and engineers are giving extremely serious consideration to the very alternatives Mr. Ramey ridicules. While Mr. Ramey scoffs at the prospects for clean power from coal and other fossil fuels, Dr. Philip Abelson, Editor of the Journal, *Science,* writes that a high national priority is the research and engineering to derive clean power from coal.* (*Science,* 1970.)

While Mr. Ramey scoffs at processes such as magneto-hydrodynamics for increasing the yield of electrical energy per unit of fuel utilized, responsible scientists such as Dr. Arthur Squires (*Science,* 1970) point up the importance of processes of this sort.**

---

*Abelson, Philip H. "Scarcity of Energy," *Science,* Vol. 169, No. 3952, Sept. 25, 1970.
**Squires, Arthur M. "Clean Power From Coal," *Science,* Vol. 169, 821-828, Aug. 28, 1970.

While Mr. Ramey scoffs at the far out aspects of solar power, highly regarded scientists point to the importance of a thorough evaluation of solar power. (*Solar Energy— Resource of the Future* by Peter E. Glaser, Ph.D., Chief of Engineering Sciences, Arthur D. Little, Inc., Acorn Park, Cambridge, Massachusetts 02140. For the National Energy Study, U.S. Dept. of Interior.)

71. **In that same speech Commissioner Ramey said:**

"I am afraid we are headed for a rather turbulent period. What does this mean, then, in terms of public relations and understanding? Obviously we are going to have to do much better in communicating with the public. We will have to show through our words, and more importantly by our actions that we indeed are performing a useful function in a responsible way with the public interest foremost in our minds."

What should one say about this statement of Commissioner Ramey?

Answer: It is a wonderful statement. Commissioner Ramey is confirming what the public already knows, namely, that the AEC has done a terribly poor job of convincing anyone that they are performing a useful function in a responsible way with the public interest foremost.

As a suggestion to Commissioner Ramey and his colleagues for their self-help, improvement programs, one might suggest one simple ingredient, candor.

72. **Commissioner Ramey further stated:**

"Nuclear power is a fact of life and I am convinced the public will reach a point in time that they will not only embrace nuclear power—they will clamor for it. So through this interim period we must retain our patience and our good humor and do the best possible job in planning and building plants properly, running them right and helping the public understand this new source of energy."

How are these statements to be viewed?

Answer: There are many "facts of life" the public never seems to clamor for. Among them are poliomyelitis, tuberculosis and cancer. Commissioner Ramey may find nuclear power as another "fact of life" that may be added to this list.

As for the statement about "helping the public understand this new source of energy," this would be an *extremely* welcome contribution on the part of the Atomic Energy Commission. A good place to start would be an honest presentation of the hazards of nuclear power, an answer to the question of the morality of going ahead with nuclear power installation when safety *cannot* be assured, and a cessation of a one-sided propaganda campaign about "the wonders of the atom."

73. **In a speech by AEC Chairman Seaborg, the following is stated (May 5, 1969):**

". . . a recent resurgence of anti-nuclear articles designed to alarm the public about the growth of nuclear power when it should be enlightened about it. Many of these articles use the effective propaganda technique known as 'stacking the deck'—the technique of taking all the detrimental, isolated facts and information about a subject, misinterpreting other factual material, adding numerous statements—taken out of context—by authorities in the field, and placing all this material in a story that gives a completely one-sided viewpoint. Specifically, every fact and statement in such a story may be true, while the article as a whole, and the conclusion it draws, may be invalid and misleading. Such dishonesty is made more harmful by the fact that these articles are written as exposes and crusades in the public interest."

What can one say about these remarks of Chairman Seaborg?

Answer: There are two important conclusions to draw from Chairman Seaborg's remarks.

(1) It is easy to understand Chairman Seaborg's dismay that people look at all the facts and come up with the "one-sided view" that they don't want nuclear power.

(2) With respect to his statement that "the article as a whole, and the conclusion it draws, may be invalid and misleading," one can only conclude that he must be referring to the numerous press releases of the U.S. Atomic Energy Commission. There are no better examples of this anywhere.

74. In the early phases of the radiation hazards controversy the nuclear power proponents denied the seriousness of the radiation hazard associated with "permissible" doses. When too many scientists began to agree that the hazard estimates were correct, the nuclear power proponents stopped the denial approach, and shifted to the claim that the standards don't mean anything since nuclear power programs would never deliver the "permissible" dose to the public.

How should the concerned public view this change in position by nuclear power proponents?

Answer: The answer is simple. If the nuclear power proponents are so sure the public will *never* be given the "permissible" doses, *they* (the nuclear proponents) should be in the *forefront* of the effort to reduce the "permissible" doses.

In a country which operates under law, one thing, and one thing alone, can guarantee that exposure to a poison will not occur, and that is to make it illegal for it to occur. It makes no sense whatever to *permit* something realized to be disastrous.

The concerned public must insist, again and again and again, that the nuclear power proponents demonstrate the sincerity of their belief that the "permissible" dose won't be reached by the only sound approach—and that approach is to *abolish the permissibility*.

318

APPENDIX II

WHEN EXPERTS DISAGREE
WHICH ONES SHALL WE BELIEVE?

It surely falls within human ability to find a way by which the scientific community, in full public view, can calmly examine together the evidence, the assumptions, and the conclusions which form the basis for assessments of radiation risks.

When the public is asked to accept an officially approved "acceptable" dose of radiation, the public is also entitled to know what risks may go with it or with any fraction of it. This approved dose is now used in determining emission standards for nuclear power plants, color television sets, and numerous other facilities and activities. Further it is used in deciding what the hazards may be from medical x-rays. If there is disagreement as to the acceptability of the dose, the public is entitled to know why.

In the hope of clarifying and perhaps resolving some disagreement regarding radiation risk-levels, I offer the following proposal. While details remain to be worked out, I see advantages to seeking public comment at this time. I welcome all suggestions.

Mike Gravel, March 1, 1971
U.S. Senate, Washington, D.C. 20510

Editor's note to the reader:

An idea like this does not "get off the ground" unless people support it. If you do, let your Senators, Representatives, and the Environmental Protection Agency hear from you.

Senate address:　　　　　　　*House of Representatives:*
Washington, D.C. 20510　　　　Washington, D.C. 20515

Environmental Protection Agency:
1129 Twentieth St., N.W.
Washington, D.C. 20460
Attention:
　　The Honorable William D. Ruckelshaus, Administrator

Proposed: A Public Radiation Debate

I: The Proposal.

II: The Need.

I. THE PROPOSAL

The word "debate" must not be interpreted literally. Proposed is an inquiry on the question:

WHAT MAY BE THE BIOLOGICAL RISK-LEVELS PER MILLION PEOPLE RECEIVING A WHOLE-BODY DOSE OF NUCLEAR RADIATION AT THE RATE OF 170 millirems PER YEAR?

In other words, what may be the health consequences of the presently permissible dose of radiation? Both somatic and genetic consequences would be considered, as would be the question of age during the exposure-period (*in utero,* childhood, adulthood).

The format would be approximately this:

As a precondition to the entire procedure, the scientists who are the principal participants or advocates in the radiation dispute would have to agree unanimously upon five scientists to be the jury. We believe this can be done.

Included in the scientific jury, it is hoped, would be at least one biologist, geneticist, statistician, public health specialist. Excluded from the scientific jury would be anyone who has publicly declared a position in this dispute, anyone who has received financial support from the Atomic Energy Commission, anyone who has been a member of professional societies related to radiation, such as the National Committee on Radiation Protection (NCRP), American Nuclear Society (ANS), American College of Radiologists, Health Physics Society, etc.

During the inquiry, all advocates and all jurors would be present, and the proceedings would be open to the public, and

320

fully recorded by a stenographer for prompt transcription.

Three sessions would be held, each approximately a month apart.

The first session: The advocates and the jurors would discuss which information, data, experiments, and papers they agree to evaluate — which animal data, which human data. The jurors would hear and participate in the arguments, and if there were disagreement, the jurors would make the final decision.

The second session: The same procedure would be used to decide what assumptions would be applied to the data — assumptions concerning public health protection, extrapolation, linearity, thresholds, dose-fractionation, dose-accumulation, cancer latency, age sensitivity, etc. Perhaps more than one constellation of assumptions would be decided upon. Again, the decision of the jurors would prevail in case of disagreement.

The third session: Each advocate would present his risk-estimates (with levels-of-confidence specified) based on applying the agreed assumptions to the agreed data.

In each session, questions and cross-examination mutually between the advocates and the jurors would proceed freely with minimal formality.

The motto would be Linus Pauling's:

"Science is the search for truth—it is not a game in which one tries to beat his opponent, to do harm to others."[1]

After the discussions, the jurors would openly deliberate and publicly announce their estimates of the BIOLOGICAL RISK-LEVELS FROM THE PRESENTLY PERMISSIBLE RADIATION DOSE (170 millirems per year whole-body exposure). Degrees of uncertainty and areas of ignorance would be described, and dissent (if any) among the jurors would be summarized.

* * * *

II. THE NEED

In January of 1970, Dr. John W. Gofman made the following statement to the Chairman of the Joint Committee on Atomic Energy:

"We urge you to nominate a jury of eminent persons, physicists, chemists, biologists, physicians, Nobel Prize winners or National Academy of Sciences members, or American Association for the Advancement of Science members, who have no atomic energy ax to grind. We urge you to serve as chairman of a debate. Dr. Tamplin and I will debate each and every facet of the evidence concerning the serious hazard of Federal Radiation Council guidelines against the entire Atomic Energy Commission staff plus anyone they can get from their 19-odd laboratories, singly, serially, or in any combination."[2]

A month later, Drs. Gofman and Tamplin wrote to Senator Edmund S. Muskie:

"Dr. Tamplin and I hereby challenge Lauriston Taylor and the entire National Committee on Radiation Protection to a complete debate, including every minute facet of the evidence, before a jury of eminent peers who have no atomic energy ax to grind, preferably in public view."[3]

Yet, unfortunately, more than a year has passed without such a debate.

The question raised by Drs. Gofman and Tamplin is simple, and it is of supreme importance to current decisions affecting our health and survival in the future:

HOW BIG A RISK (in cancer and genetically-related diseases) MAY BE ASSOCIATED WITH THE PRESENTLY PERMISSIBLE RADIATION DOSE (or "guidelines")?

The public has not received risk-estimates from the institutions officially responsible for protecting the public from radiation hazards: the Federal Radiation Council (now under the Environmental Protection Agency), the Atomic Energy Commission (AEC), the National Committee on Radiation Protection (NCRP), and the National Academy of Sciences (NAS) Advisory Committee.[4]

At the NCRP press conference on January 26, 1971, AEC spokesman, Dr. Victor Bond (who is also a member of the

NCRP) declined to express casualty figures. He instead stated: "All I can say is that we have considered exactly the same data as Drs. Gofman and Tamplin, and we come to different conclusions."

Why do equally qualified experts reach different conclusions from the same data? When experts disagree, which ones shall we believe?

These questions arise not only with radiation effects, but with cigarette smoking, drugs, pesticides, organic farming, the supersonic airplane, the Alaskan pipeline, the damming of rivers, the anti-ballistic missile, the storage of radioactive wastes, the engineering of nuclear power plants, and many other questions on which public policies have to be made and whose economic stakes are enormous. Much is at stake in addition to public health and business interests; the lofty ideals of science and democracy are also involved.

There are at least two reasons why Congressional hearings fail to cope successfully with such controversies: the hearing format does not require the experts to confront each other or to clarify the reasons for their disagreement, and members of Congress do not have the scientific training to ask all the important questions.

Expert committees are also often handicapped by deficiencies. They operate behind closed doors which exclude the public as well as the rest of the scientific community. Where conflicts of interest may be present, their findings lose credibility.[5]

Such hearings and reviews resolve little; the radiation controversy will simply rage on, unless findings can be validated in an open inquiry with all sides participating. Without such a forum, the experts will continue to roam the country independently making solo addresses, accusations, and testimonies which are not subject to critical scrutiny or even to comparison.

When scientific experts reach contradictory conclusions, there are *reasons* for the disagreement. These reasons need to be identified, and that usually requires:

    a.) Identification of the prime data which each is considering and not considering.

b.) Identification of the statistical and experimental methods used by each.

c.) Identification of the unspoken scientific premises, public health principles, and personal values underlying each expert's position.

A jury of qualified scientists could determine this information as it relates to the radiation controversy by questioning the scientists who disagree, and by listening as the disagreeing scientists question each other.

There is no reason why this examination process should be private or secret. The scientists are not the least shy about advocating their positions publicly; they can hardly claim shyness about submitting their positions publicly to the critical scrutiny of scientific peers. There is no reason, either, for the deliberations of the scientific jurors to be secret; the decency of all participants is taken for granted.

FOOTNOTES

1. *No More War,* by Linus Pauling, New York, Dodd, Mead & Co, 1958; page 209.
2. "Environmental Effects of Producing Electric Power," hearings before the Joint Committee on Atomic Energy, January and February, 1970; Part 2 (Volume I), page 1390.
3. Published in "Underground Uses of Nuclear Energy," hearings before the Senate Public Works Subcommittee on Air and Water Pollution, Nov. 18-20, 1969, Part 1, page 290.
4. When the presently permissible dose was set in 1957 at 170 millirems per year, the following risks were apparently considered acceptable: *Cancer:* According to Dr. William Mills (presently director of the Division of Biological Effects, HEW's Bureau of Radiological Health), the NCRP then estimated that 5 percent-10 percent of all cancers are caused by natural background radiation (100 millirems per year). This is an estimate also quoted by Dr. Gofman in a Nevada speech, 1965. The NRCP has set the permissible dose of *man-made* background radiation (which does not count medical radiation) at 170 millirems average for the population.
Present U.S. cancer mortality is about 320,000 deaths per year.
Genetic: the NAS Committee on the Biological Effects of Atomic Radiation (the BEAR Committee) endorsed the permissible dose, while acknowledging that the price might be a quarter of a million extra "defective children" if the parents of one generation were exposed to it every year (see Report #3 of the Federal Radiation

324

Council, May 1962. Appendix B). Now, some say that estimate was much too high, while others say it was a great underestimate.

5. In January, 1970, HEW Secretary Robert Finch directed a thorough review to be made of the permissible radiation dose. Two studies were undertaken—one by the NCRP and one by the NAS.

NCRP membership: Total 64.

About 10 are radiologists.

About 14 were also members of the BEAR Committee.

About 30 receive employment or research grants from the AEC, the Dept. of Defense, Westinghouse, or General Electric (major manufacturers of nuclear reactors).

Naturally, there is some overlap in the three categories.

NAS Radiation Committee membership: Total 20.

Twelve of the twenty are notable as follows:

7 are either employed by the AEC or have been receiving research money from the AEC.

8 are concurrently members of the NCRP.

10 were also members of the BEAR Committee.

325

APPENDIX III

NUCLEAR POWER AND ALTERNATIVES

Senator Mike Gravel has announced plans to introduce legislation which will remove preferential treatment for nuclear power plants, and give new attention to safer ways of making electricity.

One provision of the bill will be repeal of the Price-Anderson Act, which presently provides special liability limits in case of nuclear power-plant accidents; repeal of the Act may bring construction of nuclear plants to a halt. Gravel's bill will provide job insurance for the affected workers—both private and government—and indemnification for the affected businesses if construction of nuclear power plants stops.

"It is not fair—in the nuclear business, or in the defense, space, or aircraft businesses—to make people suffer for having done just what the government urged them to do," Gravel said. "On the other hand, it is not fair to the public to have allowed the construction of nuclear plants which are so potentially dangerous that a single accident might contaminate 150,000 square miles, or fifteen states the size of Maryland."

The major substance of the bill will establish an Energy-Environment Commission instead of an Atomic Energy Commission, and will provide funds to develop safe methods of generating electricity, such as clean fossil-fuel technology, magnetohydrodynamic generators, fusion, solar, and geothermal energies. Research on nuclear fission plants would also continue as one division of the new Energy-Environment Commission.

Gravel's announcement coincided with the broadcast of the NET television program, "The Advocates," in which he and Dr. John W. Gofman argued in favor of a moratorium on the construction of nuclear power plants. Senator Gravel's statement deals with the following questions:

1. **Why is it advisable to stop building nuclear power plants?**

The possibility of a major accident at one of our nuclear power plants is undeniable. One really serious accident could release as much long-lived radioactivity over the countryside as 100 Hiroshima bombs, or more. The consequences could bring this country to its knees.

2. **What might be the consequences of a major nuclear accident?**

If we use the AEC's own Brookhaven Report, we must figure the following possibilities:

Fifteen states the size of Maryland might be contaminated; agriculture restricted or forbidden; water supplies contaminated; other power plants contaminated.

Half a million people might need evacuation, fast. These radiation refugees would have no place to go, and probably no one who would want them.

Perhaps another 3½ million people might have to have their outdoor activity restricted to keep them from receiving high radiation doses.

There might be general panic, and people might demand that all the nuclear plants in the country be shut down — which would extend the economic chaos even further.

In addition, there might be 3,000 or 4,000 people dying from acute radiation overexposure.

Plus another 50,000 people dying later from radiation-induced cancer, which is a horrible way to die.

3. **Are the damage and casualty figures upper-limits on the very worst accident which could happen?**

No, the figures cited above could be significant *under-*estimates for several reasons:

a.) Nuclear plants are now being built and planned 5 times bigger than they were when the Brookhaven Report was written in 1957; that means that they produce 5 times more radioactivity per year.

b.) Because the nuclear fuel is cleaned less often now, long-lived radioactivity is given more time to accumulate inside the reactor. Therefore, at the moment of accident, a 1000-megawatt reactor may contain *more* than 5 times as much radioactivity as the 200-megawatt reactor postulated in the Brookhaven Report.

c.) The human casualties depend, of course, on how much exposure to radiation is received; if we do not succeed in evacuating up to half a million people fast enough, the casualties will go up.

d.) The Brookhaven Report postulated an accident at a small nuclear power plant located about 30 miles from a city. Huge reactors are now being built 24 miles from New York City; 12 miles from downtown Gary, Indiana; 4 miles from New London, Conn; 10 miles from Philadelphia; 5 miles from Trenton, New Jersey. Evacuation will be both more complex and more urgent.

e.) These figures also exclude all casualties caused by radiation exposure below 50 rads, which is a high dose (about 500 times more than our annual dose from natural radiation). Obviously, there *will* be additional cancers coming later from doses below 50 rads, but they are not even included in these figures. A dose of 1½ rads to a woman during pregnancy seems to increase the chance by 50% that her child will get cancer before the age of ten.

4. **Some nuclear enthusiasts refer to the Brookhaven Report as a fanciful exercise; is that true?**

The utilities take the AEC's Brookhaven Report so seriously that they have insisted on the Price-Anderson Act to limit their liability.

The AEC takes it so seriously that in 1965, the Commission admitted in writing that the consequences of accidents could be even more serious than was indicated in 1957.

If the utilities and the AEC take it seriously, we should too. If the utilities do *not* take the report seriously, then of course they will have no objection to repeal of the Price-Anderson Act.

5. **Isn't the chance nearly zero of such an accident ever occurring?**

We are told that the chances of such an accident occurring are extremely remote or negligible. That's theory, not human experience. The chance might be one chance in ten, and we would not necessarily know it yet from our accumulated experience.

The declaration of long odds—like one chance in 300 million for such an accident—is one of the most irresponsible lines being used today on the public. That's a phoney figure, both in terms of the frequency with which statistically "impossible" accidents *do* happen—like the sinking of the Titanic on its maiden voyage—and in terms of our experience so far with nuclear power plants.

We have about 100 reactor-years of experience—or some people claim 600—but we would need about 100,000 reactor-years of experience to assess odds like *one chance in 200,* if we plan 500 reactors in operation.

What were the statistical odds that the Tacoma Narrows Bridge would fall down? Surely "extremely remote." What were the odds that two airliners would collide in mid-air over the Grand Canyon? "Negligible." Or that a bomber would run into the Empire State Building?

So far, we've been lucky with a few reactors. It seems that the utilities are telling us, "Look, we haven't killed anyone yet, so give us a *chance.*"

The chance belongs to the American people, to decide whether or not they want this gamble taken with their lives and their country.

If a nuclear accident is *possible,* and they tell us it *is,* then the chance of its happening sooner is just as great as of its happening later.

6. **Is the chance of an accident growing larger or smaller?**

We've already got 20 of those radioactive power plants in operation, and any one of them *might* have an accident at any moment.

If we build more of them, the chances of accidents will increase instead of decreasing. It's not necessary to be an expert in radiation or engineering to see that humans can and do make mistakes in design, in manufacture, in construction, and in operating machines. Reactors are no exception, and we've already had some close calls with a few of them. The very act of building and operating more, allows more chances for mistakes. Especially because they are behind schedule, and rushing.

Edward Teller has warned us wisely when he said, "With the greater number of simians monkeying around with things that they do not completely understand, sooner or later a fool will prove greater than the proof even in a foolproof system."

Stopping construction will prevent the accident risks from increasing, while giving us time to consider such possibilities as building all nuclear reactors underground, or developing other kinds of power.

7. **Why do utilities advertise nuclear power plants as safe?**

Obviously, there is a puzzling contradiction between the utilities' advertisements which claim radioactive power plants are wonderfully safe, and the utilities' testimony to Congress that they would not build them unless Congress relieved them of almost all financial responsibility for accidents. If "nukes" are as safe as they claim, why do they worry about financial responsibility for *accidents?*

If the utilities won't even risk their *dollars* on the safety of nuclear power plants, why should the people have to risk their *lives?*

We should not wait for the Price-Anderson Act to expire in 1977. Repeal *now* is a minimum objective.

8. **Is a move against nuclear electricity a move against progress?**

Progress in technology might be defined as something which *enhances* human health and survival. The *one* technology which has the ability to pollute this planet permanently is hard to consider as *progress*.

We've spent billions on nuclear research, we're buying nightmares for generations to come, and for what?

We end up with *another* way to boil water. That's all that a nuclear reactor accomplishes. It boils water, which produces electricity very *in*efficiently, and it also produces radioactive garbage to the tune of about 1,000 Hiroshima bombs-worth a year.

Is that human progress?

One of the main ingredients of the nuclear power program is plutonium-239, which lingers radioactively for 240,000 years. Other kinds of radioactive waste last hundreds of years.

Who needs it?

9. **We already have power shortages; won't the lights go out for sure if we have a nuclear moratorium?**

The lights won't go out because of a *nuclear moratorium,* although they may go out due to *other* foul-ups.

For instance, in 1969, the utilities spent about $320 million on advertising to increase consumption of electricity—and only $41 million on research and development of ways to generate it. *No wonder* we have a power shortage.

Brown-outs and black-outs won't be because of a moratorium. In fact, we will hardly miss nuclear power at all, which is only one percent to two percent of our power supply now.

10. **Will a nuclear moratorium cause chaos and unemployment in the power industry?**

We have chaos *now,* even without a nuclear moratorium. Every analysis of the power shortage refers to bad planning and miscalculation on the part of the industry.

Of course we can expect to hear wailing and cries of "We can't deliver the power," even from the coal operators who are sitting on a 400-year supply of coal. We also hear the can-not-do cries from the automobile industry about clean cars. Do you believe them?

This country could declare a moratorium tomorrow—and we might even hear a sigh of relief from some worried people *inside* the nuclear establishment—provided we insured jobs and offered indemnification. After all, we pay landowners billions of dollars every year *not* to grow crops; we can certainly afford to pay people *not* to build radioactive machines which could contaminate an area from New York City to Richmond, Virginia.

If we take care of the financial hardships of a moratorium, arguments in favor of nuclear electricity may lose some of their frenzy.

My proposal for repeal of the Price-Anderson Act is part of legislation which includes establishment of an Energy-Environment Commission. There will be *more* energy business and *more* energy employment, not less, because the total energy effort in this country needs to be far *greater* than it is.

11. What alternatives are there to nuclear electricity?

In California alone, there seems to be geothermal steam in the ground equal to the power of 20 big radioactive power plants. Geothermal steam is not only perfectly clean—it's also safe. There is lots of it in the west. Enough for several hundred years.

In addition, this country has enough coal to provide electricity for the next 400 years—the present shortage is both temporary and artificial.

In August 1970, the Vice President of the National Coal Association testified under oath that the mine operators can go just as fast mining coal, as the power demands can grow. But it won't be sensible to open coal mines unless the utilities offer long-term contracts.

The lead story in the magazine *Science News,* January

332

30, 1971, is "Coal's Road Toward Acceptability." It makes some important points about our ability to make coal a clean fuel, and our ability to restore land ruined by strip-mining.

However, fossil-fuels should not be considered a long-term solution for generating electricity. It is short-sighted to waste the planet's exhaustible natural resources by burning them, when all we have to do is tap into the *in*exhaustible sources of energy like water, wind, geothermal heat, and the sun.

12. Will a nuclear moratorium increase air pollution by forcing us back into coal and other dirty fossil-fuels?

No. And it's dishonest to tell the public that the only choice is between clean "nukes" and dirty coal.

I favor forcing the coal-plants to clean up, and *fast,* because even without a nuclear moratorium, we must depend on fossil-fuels to make most of our electricity for the next 20 or 30 years. Dirty plants are an intolerable abuse of public health, so it's a necessity that we clean up the old plants, and build the new ones *clean.*

It can be done. In fact, equipment to do most of it is already available. We need to see that the utilities buy and use it.

Of course the equipment may not work perfectly at first. But it's far safer to take some chances with unproven fossil-fuel equipment than with unproven nuclear equipment. That's obvious.

13. Will a nuclear moratorium just delay the nuclear plants we'll need sooner or later anyway?

No, because nuclear plants are not inevitable. Perhaps some day we will find a way to make them truly accident-proof; additional safety research is urgently needed *now* according to the AEC's own Advisory Committee on Reactor Safeguards.

We need time to do that research *before* we build more plants, and also to give the country a chance to look at other new ways to make electricity—ways which are not tied to potential catastrophe.

14. Is it possible to generate electricity without pollution?

The answer is probably yes. No one knows, because we haven't begun really trying yet.

Consider some of the possibilities we are presently neglecting, even though we know they are *real* possibilities:

Magnetohydrodynamic generators (MHD), which would *contain* the fossil-fuel pollutants.

Fusion power, which could provide energy for the entire world from seawater—and lower ocean levels by far less than one-thousandth of an inch over the next million years.

Geothermal energy, which is pollution-free and accessible anywhere by drilling 5 to 10 miles down.

Solar energy, whose energy supply both in the United States and in the world, is far greater than any possible needs; it may already be technically possible to recover about 4 trillion kilowatt-hours in electrical energy per year from Death Valley alone; the use of solar collectors in orbit is another clear possibility.

We should consider wind and tidal power too. There is such a fabulous amount of energy renewing itself naturally on earth that, if man tapped only a tiny percent of it, he could probably make all the electricity he needs *without* poisoning the planet or disturbing its natural rhythms.

The real question to decide during a nuclear halt is:

Do we take our chances with some of the gentle possibilities, or do we rush into a commitment to the *one* technology which may end up contaminating this planet permanently?

What this country needs urgently is an Energy Commission, instead of an *Atomic* Energy Commission.

334

15. **How many years away are these alternatives to nuclear power?**

Obviously, that depends on how much effort we start putting into them. Therefore, the following figures are just estimates; some of the possibilities may *never* be practical, but we need only one or two to work out.

Fossil-fuels: Removal of sulfur pollutants: now. Removal of nitrogen, mercury, and radioactive particles: 5 years.

MHD generators: Apparently Russia already has a small pilot plant working. AVCO Everett Lab in Massachusetts, which demonstrated MHD feasibility more than 10 years ago, is now designing a 50-megawatt commercial generator with Con Ed of New York, Boston Edison, and some northeast utilities.

Fusion: Feasibility might be proven within the next five years, followed by demonstration plants in the 1980's, and commercial operation before the end of the century.

Geothermal energy: Natural geothermal steam is practical now; it's already producing electricity commercially in California, Italy, Mexico, Japan, New Zealand, and Russia. In order to make geothermal energy available just about everywhere on earth, we need to develop deep drilling techniques, which might take 10 years.

Solar energy: Land-based techniques for recovery, storage, and transmission of solar energy are theoretically possible already; engineering large-scale projects might take 10 years; techniques to collect the energy from orbiting stations are probably 25 years away.

16. **Is the government investing equally in all the alternatives?**

Unfortunately, for years we've been putting about 83 percent of the federal energy-research dollar into radioactive power plants, and almost nothing for the other possibilities.

Last year, the government spent approximately

$255,000,000 on developing radioactive nuclear power plants.

30,000,000 on developing fusion power.

300,000 on developing MHD generators.

zero   on developing geothermal technology.

zero   on developing solar energy.

In other words, we spent less on developing non-radioactive sources of power than we spent on two 747 airliners.

In fact, when we take inflation into account, the effort in fusion will *decrease* again this year under the AEC's plans. The AEC is in charge of both fusion *and* fission (radioactive power plants).

17. Would the cost of household electricity have to go up to pay for this energy development?

No. For one thing, the $320,000,000 which is spent per year now on advertising electricity could be spent on energy research instead.

Then the rate-structure could be reversed, so that the more electricity you use, the more it costs you. Right now, the more you use, the *less* it costs you per kilowatt-hour. Obviously, if we have a power problem, we should not reward people for using *more* electricity, and punish the little person because he uses *less* of it.

In addition, a federal tax on the electrical bills of the big industrial users would pay for the plan without raising household bills. This would be consistent with the long-time policy of this country to avoid regressive taxes and regressive charges, which hit people who can least afford them.

We've got to face a fact, however. Probably *no one* has been paying the true cost of electricity on his bill. We have been paying its true cost instead in pollution and in medical bills.

Nuclear power, which was supposed to be "too cheap

336

to meter," has turned out to be the most expensive power we have, even with all the hidden government subsidies. Let's not make any more foolish predictions about the cost of electricity. Safe, clean alternatives may ultimately cost less *or* more. There is a good chance they will cost less.

18. How long might it be before an Energy-Environment Commission is usefully in operation?

There will be lots of hassles over jurisdiction, over powers, over sources of revenue, over allocation of revenue, over new kinds of administrative organization designed to avoid bureaucratic inertia, and so forth. I expect controversy, and also I expect improvements to be made on my bill.

Invariably, these things take more time than necessary . . . which makes it important to start *now*.

19. Is there anything which can be done in the meantime?

Since the energy problem is so urgent, we ought to push for bills *this* session to get initial funding for solar energy and geothermal energy, plus more for fusion than is now in the AEC's budget. If we wait for a perfect new agency to come into existence, we'll lose more time. We can get started *this* year by pushing for programs under existing agencies.

Obviously, it will make a lot more sense to have an Energy-Environment Commission supervising our energy efforts, but we should not use its non-existence to postpone our efforts. Nor should we just study the problem, which is obvious. We need action toward solutions.

Do you realize what it will mean to this earth if we *don't* start action? We'll find ourselves toe-to-toe due to the population explosion, with only primitive energy systems to provide a tolerably human standard of living for billions of people. We would pollute or contaminate this planet beyond tolerance if we had to depend on today's energy technology.

Therefore, it's really a very modest proposal to start funding let's say 2,000 people—out of a labor force of about 70 million Americans—whose *job* will be to make solar energy practical. The same is true for deep geothermal wells. And the fusion budget should be tripled at least.

Some will say, "But the program can't absorb the money," but the can-not-do attitude is nothing new. Can anyone really argue that putting 2,000 new brains, with new ideas, on the problems of fusion would be a waste of money?

That's the kind of immediate effort I'm supporting. Unless we put *effort* into solving energy needs, it's insincere to say that it's electricity *vs.* the environment, or any of the other false choices offered us.

Senator Mike Gravel
February 15, 1971

APPENDIX IV

Commercial Nuclear Power Reactors in the United States

Site	Plant Name	Capacity Net kW(e)	Utility	Commercial Operation
ALABAMA				
Decatur	Browns Ferry Nuclear Power Plant: Unit 1	1,065,000	Tennessee Valley Authority	1974
Decatur	Browns Ferry Nuclear Power Plant: Unit 2	1,065,000	Tennessee Valley Authority	1975
Decatur	Browns Ferry Nuclear Power Plant: Unit 3	1,065,000	Tennessee Valley Authority	1977
Dothan	Joseph M. Farley Nuclear Plant: Unit 1	829,000	Alabama Power Co.	1977
Dothan	Joseph M. Farley Nuclear Plant: Unit 2	820,000	Alabama Power Co.	1980
Scottsboro	Bellefonte Nuclear Plant: Unit 1	1,213,000	Tennessee Valley Authority	1981
Scottsboro	Bellefonte Nuclear Plant: Unit 2	1,213,000	Tennessee Valley Authority	1982
ARIZONA				
Wintersburg	Palo Verde Nuclear Generating Station: Unit 1	1,270,000	Arizona Public Service	1982
Wintersburg	Palo Verde Nuclear Generating Station: Unit 2	1,270,000	Arizona Public Service	1984
Wintersburg	Palo Verde Nuclear Generating Station: Unit 3	1,270,000	Arizona Public Service	1986
Wintersburg	Palo Verde Nuclear Generating Station: Unit 4	1,270,000	Arizona Public Service	1988
Wintersburg	Palo Verde Nuclear Generating Station: Unit 5	1,270,000	Arizona Public Service	1990
ARKANSAS				
Russellville	Arkansas Nuclear One: Unit 1	850,000	Arkansas Power & Light Co.	1974
Russellville	Arkansas Nuclear One: Unit 2	912,000	Arkansas Power & Light Co.	1979
CALIFORNIA				
Eureka	Humboldt Bay Power Plant: Unit 3	63,000	Pacific Gas & Electric Co.	1963
San Clemente	San Onofre Nuclear Generating Station: Unit 1	436,000	So. Calif. Ed. & San Diego Gas & El. Co.	1968

Site	Plant Name	Capacity Net kW(e)	Utility	Commercial Operation
(California, cont.)				
San Clemente	San Onofre Nuclear Generating Station: Unit 2	1,100,000	So. Calif. Ed. & San Diego Gas & El. Co.	1981
San Clemente	San Onofre Nuclear Generating Station: Unit 3	1,100,000	So. Calif. Ed. & San Diego Gas & El. Co.	1983
Diablo Canyon	Diablo Canyon Nuclear Power Plant: Unit 1	1,084,000	Pacific Gas & Electric Co.	1979
Diablo Canyon	Diablo Canyon Nuclear Power Plant: Unit 2	1,106,000	Pacific Gas & Electric Co.	1979
Clay Station	Rancho Seco Nuclear Generating Station	918,000	Sacramento Municipal Utility District	1975
Site not selected	Unit 1	1,200,000	Pacific Gas & Electric Co.	Indef.
Site not selected	Unit 2	1,200,000	Pacific Gas & Electric Co.	Indef.
COLORADO				
Platteville	Ft. St. Vrain Nuclear Generating Station	330,000	Public Service Co. of Colorado	1978
CONNECTICUT				
Haddam Neck	Haddam Neck Plant	575,000	Conn. Yankee Atomic Power Co.	1968
Waterford	Millstone Nuclear Power Station: Unit 1	660,000	Northeast Nuclear Energy Co.	1971
Waterford	Millstone Nuclear Power Station: Unit 2	830,000	Northeast Nuclear Energy Co.	1975
Waterford	Millstone Nuclear Power Station: Unit 3	1,156,000	Northeast Nuclear Energy Co.	1986
FLORIDA				
Florida City	Turkey Point Station: Unit 3	693,000	Florida Power & Light Co.	1972
Florida City	Turkey Point Station: Unit 4	693,000	Florida Power & Light Co.	1973
Red Cove	Crystal River Plant: Unit 3	825,000	Florida Power Corporation	1977
Ft. Pierce	St. Lucie Plant: Unit 1	802,000	Florida Power & Light Co.	1976
Ft. Pierce	St. Lucie Plant: Unit 2	810,000	Florida Power & Light Co.	1983
GEORGIA				
Baxley	Edwin I. Hatch Nuclear Plant: Unit 1	786,000	Georgia Power Co.	1975
Baxley	Edwin I. Hatch Nuclear Plant: Unit 2	795,000	Georgia Power Co.	1978

Site	Plant Name	Capacity Net kW(e)	Utility	Commercial Operation
(Georgia, cont.)				
Waynesboro	Alvin W. Vogtle, Jr. Plant: Unit 1	1,110,000	Georgia Power Co.	1984
Waynesboro	Alvin W. Vogtle, Jr. Plant: Unit 2	1,110,000	Georgia Power Co.	1987
ILLINOIS				
Morris	Dresden Nuclear Power Station: Unit 1	200,000	Commonwealth Edison Co.	1960
Morris	Dresden Nuclear Power Station: Unit 2	794,000	Commonwealth Edison Co.	1970
Morris	Dresden Nuclear Power Station: Unit 3	794,000	Commonwealth Edison Co.	1971
Zion	Zion Nuclear Plant: Unit 1	1,040,000	Commonwealth Edison Co.	1973
Zion	Zion Nuclear Plant: Unit 2	1,040,000	Commonwealth Edison Co.	1974
Cordova	Quad-Cities Station: Unit 1	789,000	Comm. Ed. Co. Ia.-Ill. Gas & Elec. Co.	1973
Cordova	Quad-Cities Station: Unit 2	789,000	Comm. Ed. Co. Ia.-Ill. Gas & Elec. Co.	1973
Seneca	LaSalle County Nuclear Station: Unit 1	1,078,000	Commonwealth Edison Co.	1979
Seneca	LaSalle County Nuclear Station: Unit 2	1,078,000	Commonwealth Edison Co.	1980
Byron	Byron Station: Unit 1	1,120,000	Commonwealth Edison Co.	1981
Byron	Byron Station: Unit 2	1,120,000	Commonwealth Edison Co.	1982
Braidwood	Braidwood: Unit 1	1,120,000	Commonwealth Edison Co.	1981
Braidwood	Braidwood: Unit 2	1,120,000	Commonwealth Edison Co.	1982
Clinton	Clinton Nuclear Power Plant: Unit 1	933,400	Illinois Power Co.	1982
Clinton	Clinton Nuclear Power Plant: Unit 2	933,400	Illinois Power Co.	1988
Savanna	Carroll County Station: Unit 1	1,120,000	Commonwealth Edison Co.	1987
Savanna	Carroll County Station: Unit 2	1,120,000	Commonwealth Edison Co.	1988
INDIANA				
Westchester	Bailly Generating Station	643,000	Northern Indiana Public Service Co.	1984
Madison	Marble Hill Nuclear Power Station: Unit 1	1,130,000	Public Service Indiana	1982
Madison	Marble Hill Nuclear Power Station: Unit 2	1,130,000	Public Service Indiana	1984

Site	Plant Name	Capacity Net kW(e)	Utility	Commercial Operation
IOWA				
Palo	Duane Arnold Energy Center: Unit 1	538,000	Iowa Electric Light and Power Co.	1975
Vandalia	Vandalia Nuclear Project	1,270,000	Iowa Power & Light Co.	Indef.
KANSAS				
Burlington	Wolf Creek Generating Station: Unit 1	1,150,000	Kansas Gas & Electric—Kansas City P & L	1983
LOUISIANA				
Taft	Waterford Generating Station: Unit 3	1,113,000	Louisiana Power & Light Co.	1981
St. Francisville	River Bend Station: Unit 1	934,000	Gulf States Utilities Co.	1984
St. Francisville	River Bend Station: Unit 2	934,000	Gulf States Utilities Co.	Indef.
MAINE				
Wiscasset	Maine Yankee Atomic Power Plant	790,000	Maine Yankee Atomic Power Co.	1972
MARYLAND				
Lusby	Calvert Cliffs Nuclear Power Plant: Unit 1	845,000	Baltimore Gas and Electric Co.	1975
Lusby	Calvert Cliffs Nuclear Power Plant: Unit 2	845,000	Baltimore Gas and Electric Co.	1977
MASSACHUSETTS				
Rowe	Yankee Nuclear Power Station	175,000	Yankee Atomic Electric Co.	1961
Plymouth	Pilgrim Station: Unit 1	655,000	Boston Edison Co.	1972
Plymouth	Pilgrim Station: Unit 2	1,150,000	Boston Edison Co.	1985
Montague	Montague: Unit 1	1,150,000	Northeast Utilities	Indef.
Montague	Montague: Unit 2	1,150,000	Northeast Utilities	Indef.

Site	Plant Name	Capacity Net kW(e)	Utility	Commercial Operation
MICHIGAN				
Big Rock Point	Big Rock Point Nuclear Plant	72,000	Consumers Power Co.	1963
South Haven	Palisades Nuclear Power Station	805,000	Consumers Power Co.	1971
Newport	Enrico Fermi Atomic Power Plant: Unit 2	1,093,000	Detroit Edison Co.	1980
Bridgman	Donald C. Cook Plant: Unit 1	1,054,000	Indiana & Michigan Electric Co.	1975
Bridgman	Donald C. Cook Plant: Unit 2	1,100,000	Indiana & Michigan Electric Co.	1978
Midland	Midland Nuclear Power Plant: Unit 1	460,000	Consumers Power Co.	1982
Midland	Midland Nuclear Power Plant: Unit 2	811,000	Consumers Power Co.	1981
St. Clair County	Greenwood: Unit 2	1,200,000	Detroit Edison Co.	1987
St. Clair County	Greenwood: Unit 3	1,200,000	Detroit Edison Co.	1989
MINNESOTA				
Monticello	Monticello Nuclear Generating Plant	545,000	Northern States Power Co.	1971
Red Wing	Prairie Island Nuclear Generating Plant: Unit 1	530,000	Northern States Power Co.	1973
Red Wing	Prairie Island Nuclear Generating Plant: Unit 2	530,000	Northern States Power Co.	1974
MISSISSIPPI				
Corinth	Yellow Creek: Unit 1	1,285,000	Tennessee Valley Authority	1985
Corinth	Yellow Creek: Unit 2	1,285,000	Tennessee Valley Authority	1986
Port Gibson	Grand Gulf Nuclear Station: Unit 1	1,250,000	Mississippi Power & Light Co.	1981
Port Gibson	Grand Gulf Nuclear Station: Unit 2	1,250,000	Mississippi Power & Light Co.	1984
MISSOURI				
Fulton	Callaway Plant: Unit 1	1,120,000	Union Electric Co.	1982
Fulton	Callaway Plant: Unit 2	1,120,000	Union Electric Co.	1987

Site	Plant Name	Capacity Net kW(e)	Utility	Commercial Operation
NEBRASKA				
Fort Calhoun	Ft. Calhoun Station: Unit 1	457,000	Omaha Public Power District	1973
Brownville	Cooper Nuclear Station	778,000	Nebraska Pub. Power. Dist. & Iowa P&L Co.	1974
NEW HAMPSHIRE				
Seabrook	Seabrook Nuclear Station: Unit 1	1,200,000	Public Service of N.H.	1983
Seabrook	Seabrook Nuclear Station: Unit 2	1,200,000	Public Service of N.H.	1985
NEW JERSEY				
Toms River	Oyster Creek Nuclear Power Plant: Unit 1	650,000	Jersey Central Power & Light Co.	1969
Forked River	Forked River Generating Station: Unit 1	1,070,000	Jersey Central Power & Light Co.	1984
Salem	Salem Nuclear Generating Station: Unit 1	1,090,000	Public Service Electric and Gas, N.J.	1977
Salem	Salem Nuclear Generating Station: Unit 1	1,115,000	Public Service Electric and Gas, N.J.	1979
Salem	Hope Creek Generating Station: Unit 1	1,067,000	Public Service Electric and Gas, N.J.	1984
Salem	Hope Creek Generating Station: Unit 2	1,067,000	Public Service Electric and Gas, N.J.	1986
NEW YORK				
Buchanan	Indian Point Station: Unit 1	265,000	Consolidated Edison Co.	1962
Buchanan	Indian Point Station: Unit 2	873,000	Consolidated Edison Co.	1973
Buchanan	Indian Point Station: Unit 3	965,000	Power Authority of State of N.Y.	1976
Scriba	Nine Mile Point Nuclear Station: Unit 1	610,000	Niagara Mohawk Power Corp.	1969
Scriba	Nine Mile Point Nuclear Station: Unit 2	1,099,800	Niagara Mohawk Power Corp.	1983
Ontario	R.E. Ginna Nuclear Power Plant: Unit 1	490,000	Rochester Gas & Electric Corp.	1970
Brookhaven	Shoreham Nuclear Power Station	819,000	Long Island Lighting Co.	1980
Scriba	James A. FitzPatrick Nuclear Power Plant	821,000	Power Authority of State of N.Y.	1975
Cementon	Greene County Nuclear Power Plant	1,212,000	Power Authority of State of N.Y.	1986

Site	Plant Name	Capacity Net kW(e)	Utility	Commercial Operation
(New York, cont.)				
Jamesport	Jamesport 1	1,150,000	Long Island Lighting Co.	1988
Jamesport	Jamesport 2	1,150,000	Long Island Lighting Co.	1990
Oswego	Sterling Nuclear: Unit 1	1,150,000	Rochester Gas & Electric Co.	1988
New Haven	New York State Electric & Gas: Unit 1	1,250,000	New York State Electric & Gas Co.	1994
New Haven	New York State Electric & Gas: Unit 2	1,250,000	New York State Electric & Gas Co.	1996
NORTH CAROLINA				
Southport	Brunswick Steam Electric Plant: Unit 1	821,000	Carolina Power and Light Co.	1977
Southport	Brunswick Steam Electric Plant: Unit 2	821,000	Carolina Power and Light Co.	1975
Cowans Ford Dam	Wm. B. McGuire Nuclear Station: Unit 1	1,180,000	Duke Power Co.	1979
Cowans Ford Dam	Wm. B. McGuire Nuclear Station: Unit 1	1,180,000	Duke Power Co.	1981
Bonsal	Shearon Harris Plant: Unit 1	900,000	Carolina Power and Light Co.	1984
Bonsal	Shearon Harris Plant: Unit 2	900,000	Carolina Power and Light Co.	1986
Bonsal	Shearon Harris Plant: Unit 3	900,000	Carolina Power and Light Co.	1990
Bonsal	Shearon Harris Plant: Unit 4	900,000	Carolina Power and Light Co.	1988
Davie County	Perkins Nuclear Station: Unit 1	1,280,000	Duke Power Co.	1988
Davie County	Perkins Nuclear Station: Unit 2	1,280,000	Duke Power Co.	1991
Davie County	Perkins Nuclear Station: Unit 3	1,280,000	Duke Power Co.	1993
Site not selected		1,150,000	Carolina Power and Light Co.	Indef.
Site not selected		1,150,000	Carolina Power and Light Co.	1989
OHIO				
Berlin Heights	Erie Nuclear Plant: Unit 1	1,260,000	Ohio Edison Co.	1986
Berlin Heights	Erie Nuclear Plant: Unit 2	1,260,000	Ohio Edison Co.	1988
Oak Harbor	Davis-Besse Nuclear Power Station: Unit 1	906,000	Toledo Edison-Cleveland El. Illum. Co.	1977
Oak Harbor	Davis-Besse Nuclear Power Station: Unit 2	906,000	Toledo Edison-Cleveland, El. Illum. Co.	1985

Site	Plant Name	Capacity Net kW(e)	Utility	Commercial Operation
(Ohio, cont.)				
Oak Harbor	Davis-Besse Nuclear Power Station: Unit 3	906,000	Toledo Edison-Cleveland, El. Illum. Co.	1987
Perry	Perry Nuclear Power Plant: Unit 1	1,205,000	Cleveland Electric Illuminating Co.	1981
Perry	Perry Nuclear Power Plant: Unit 2	1,205,000	Cleveland Electric Illuminating Co.	1983
Moscow	Wm. H. Zimmer Nuclear Power Station: Unit 1	810,000	Cincinnati Gas & Electric Co.	1980
OKLAHOMA				
Inola	Black Fox Nuclear Station: Unit 1	1,150,000	Public Service of Oklahoma	1984
Inola	Black Fox Nuclear Station: Unit 2	1,150,000	Public Service of Oklahoma	1986
OREGON				
Prescott	Trojan Nuclear Plant: Unit 1	1,130,000	Portland General Electric Co.	1976
Arlington	Pebble Springs Nuclear Plant: Unit 1	1,260,000	Portland General Electric Co.	1987
Arlington	Pebble Springs Nuclear Plant: Unit 2	1,260,000	Portland General Electric Co.	1989
PENNSYLVANIA				
Peach Bottom	Peach Bottom Atomic Power Station: Unit 2	1,065,000	Philadelphia Electric Co.	1974
Peach Bottom	Peach Bottom Atomic Power Station: Unit 3	1,065,000	Philadelphia Electric Co.	1974
Pottstown	Limerick Generating Station: Unit 1	1,065,000	Philadelphia Electric Co.	1983
Pottstown	Limerick Generating Station: Unit 2	1,065,000	Philadelphia Electric Co.	1985
Shippingport	Shippingport Atomic Power Station	60,000	Department of Energy	1957
Shippingport	Beaver Valley Power Station: Unit 1	852,000	Duquesne Light Co.-Ohio Edison Co.	1976
Shippingport	Beaver Valley Power Station: Unit 2	833,000	Duquesne Light Co.-Ohio Edison Co.	1984
Middletown	Three Mile Island Nuclear Station: Unit 1	819,000	Metropolitan Edison Co.	1974
Middletown	Three Mile Island Nuclear Station: Unit 2	906,000	Jersey Central Power & Light Co.	1979

Site	Plant Name	Capacity Net kW(e)	Utility	Commercial Operation
(Pennsylvania, cont.)				
Berwick	Susquehanna Steam Electric Station: Unit 1	1,050,000	Pennsylvania Power & Light Co.	1981
Berwick	Susquehanna Steam Electric Station: Unit 2	1,050,000	Pennsylvania Power & Light Co.	1982
RHODE ISLAND				
Charlestown	New England Power (NEP): Unit 1	1,150,000	New England Power Co.	1987
Charlestown	New England Power (NEP): Unit 2	1,150,000	New England Power Co.	1989
SOUTH CAROLINA				
Hartsville	H.B. Robinson S.E. Plant: Unit 2	700,000	Carolina Power and Light Co.	1971
Seneca	Oconee Nuclear Station: Unit 1	887,000	Duke Power Co.	1973
Seneca	Oconee Nuclear Station: Unit 2	887,000	Duke Power Co.	1974
Seneca	Oconee Nuclear Station: Unit 3	887,000	Duke Power Co.	1974
Broad River	Virgil C. Summer Nuclear Station: Unit 1	900,000	South Carolina Electric and Gas Co.	1983
Lake Wylie	Catawba Nuclear Station: Unit 1	1,145,000	Duke Power Co.	1981
Lake Wylie	Catawba Nuclear Statoin: Unit 2	1,145,000	Duke Power Co.	1983
Cherokee County	Cherokee Nuclear Station: Unit 1	1,280,000	Duke Power Co.	1985
Cherokee County	Cherokee Nuclear Station: Unit 2	1,280,000	Duke Power Co.	1987
Cherokee County	Cherokee Nuclear Station: Unit 3	1,280,000	Duke Power Co.	1989
TENNESSEE				
Daisy	Sequoyah Nuclear Power Plant: Unit 1	1,148,000	Tennessee Valley Authority	1979
Daisy	Sequoyah Nuclear Power Plant: Unit 2	1,148,000	Tennessee Valley Authority	1980
Spring City	Watts Bar Nuclear Plant: Unit 1	1,177,000	Tennessee Valley Authority	1979
Spring City	Watts Bar Nuclear Plant: Unit 2	1,177,000	Tennessee Valley Authority	1980
Oak Ridge	Clinch River Breeder Reactor Plant	350,000	Department of Energy	1988
Hartsville	A, Unit 1	1,233,000	Tennessee Valley Authority	1983

Site	Plant Name	Capacity Net kW(e)	Utility	Commercial Operation
(Tennessee, cont.)				
Hartsville	A, Unit 2	1,233,000	Tennessee Valley Authority	1984
Hartsville	B, Unit 1	1,233,000	Tennessee Valley Authority	1983
Hartsville	B, Unit 2	1,233,000	Tennessee Valley Authority	1984
Kingsport	Phipps Bend, Unit 1	1,233,000	Tennessee Valley Authority	1984
Kingsport	Phipps Bend, Unit 2	1,233,000	Tennessee Valley Authority	1985
TEXAS				
Glen Rose	Comanche Peak Steam Electric Station: Unit 1	1,111,000	Texas Utilities Services, Inc.	1981
Glen Rose	Comanche Peak Steam Electric Station: Unit 2	1,111,000	Texas Utilities Services, Inc.	1983
Wallis	Allens Creek: Unit 1	1,150,000	Houston Lighting & Power Co.	1985
Matagorda County	South Texas: Unit 1	1,250,000	Central Power & Lt.-Houston Lt. & Power	1980
Matagorda County	South Texas: Unit 2,	1,250,000	Central Power & Lt.-Houston Lt. & Power	1982
VERMONT				
Vernon	Vermont Yankee Generating Station	514,000	Vermont Yankee Nuclear Power Corp.	1972
VIRGINIA				
Gravel Neck	Surry Power Station: Unit 1	822,000	Virginia Electric & Power Company	1972
Gravel Neck	Surry Power Station: Unit 2	822,000	Virginia Electric & Power Company	1973
Mineral	North Anna Power Station: Unit 1	907,000	Virginia Electric & Power Company	1978
Mineral	North Anna Power Station: Unit 2	907,000	Virginia Electric & Power Company	1979
Mineral	North Anna Power Station: Unit 3	907,000	Virginia Electric & Power Company	1983
Mineral	North Anna Power Station: Unit 4	907,000	Virginia Electric & Power Company	1984
WASHINGTON				
Richland	N. Reactor/WPPSS Steam	850,000	Department of Energy	1966

Site	Plant Name	Capacity Net kW(e)	Utility	Commercial Operation
(Washington, cont.)				
Richland	WPPSS No. 1	1,218,000	Washington Public Power Supply System	1982
Richland	WPPSS No. 2	1,100,000	Washington Public Power Supply System	1980
Satsop	WPPSS No. 3	1,242,000	Washington Public Power Supply System	1984
Richland	WPPSS No. 4	1,218,000	Washington Public Power Supply System	1984
Satsop	WPPSS No. 5	1,242,000	Washington Public Power Supply System	1985
Sedro Woolley	Skagit Nuclear Project: Unit 1	1,277,000	Puget Sound Power & Light	1986
Sedro Woolley	Skagit Nuclear Project: Unit 2	1,277,000	Puget Sound Power & Light	1988
WISCONSIN				
La Cross	La Cross (Genoa) Nuclear Generating Station	50,000	Dairyland Power Coorporation	1969
Two Creeks	Point Beach Nuclear Plant: Unit 1	497,000	Wisconsin Michigan Power Co.	1970
Two Creeks	Point Beach Nuclear Plant: Unit 2	497,000	Wisconsin Michigan Power Co.	1972
Carlton	Kewaunee Nuclear Power Plant: Unit 1	535,000	Wisconsin Public Service Corp.	1974
Haven	Haven Nuclear Plant: Unit 1	900,000	Wisconsin Electric Power Co.	1987
Haven	Haven Nuclear Plant: Unit 2	900,000	Wisconsin Electric Power Co.	Indef.
Durand	Tyrone Energy Park: Unit 1	1,100,000	Northern States Power	1985

Index

Bibliography

* Books for young readers

*Bader, Bonnie. *Who Was Martin Luther King, Jr.?* New York: Penguin Workshop, 2008.

Bagley, Edythe Scott, with Joe Hilley. *Desert Rose: The Life and Legacy of Coretta Scott King.* Tuscaloosa: University of Alabama Press, 2012.

King, Coretta Scott. *My Life with Martin Luther King, Jr.* New York: Henry Holt and Company, 1993.

*Krull, Kathleen. *What Was the March on Washington?* New York: Penguin Workshop, 2013.

*Medearis, Angela Shelf. *Dare to Dream: Coretta Scott King and the Civil Rights Movement.* New York: Puffin Books, 1994.

Vivian, Octavia B. *Coretta: The Story of Coretta Scott King.* Minneapolis: Fortress Press, 2006.

Timeline of the World

1927	Charles Lindbergh makes first nonstop solo flight across the Atlantic Ocean
	The Jazz Singer premieres, ushering in "talking" movies
1939	World War II begins
1947	Jackie Robinson becomes first black player on a Major League Baseball team
1954	Supreme Court rules school segregation unconstitutional in *Brown v. Board of Education*
1955	Ray Kroc opens the first McDonald's in Des Plaines, Illinois
1960	Seventeen African countries gain independence from European rule
1962	Astronaut John Glenn is the first American to orbit Earth
1963	President John F. Kennedy is assassinated on November 22
1965	President Lyndon Johnson signs the Voting Rights Act of 1965
	First US combat troops sent to Vietnam
1968	The computer mouse, invented by Douglas C. Engelbart, and the world's first jumbo jet, the 747, are both introduced
1985	Wreck of the *Titanic* is discovered
	Michael Jordan named NBA Rookie of the Year
2006	Construction begins on One World Trade Center in New York City

Timeline of Coretta Scott King's Life

1927	Coretta Scott King is born April 27
1945	Goes to Antioch College in Ohio
1953	Marries Martin Luther King Jr.
1954	Moves to Montgomery, Alabama
1955	First child, Yolanda, is born
	Helps start the Montgomery Bus Boycott
1957	Second child, Martin Luther King III, is born
1959	Moves to Atlanta, where Martin is a pastor at Ebenezer Baptist Church
1961	Third child, Dexter, is born
1962	Serves as delegate at an international peace conference in Switzerland
1963	Fourth child, Bernice, is born
	Marches at the March on Washington
1964	Holds first Freedom Concert in New York City
1965	Joins the Selma March
1968	Leads Memphis march in Martin's place after he is killed
	Establishes Martin Luther King Jr. Center for Nonviolent Social Change
1985	Is arrested in Washington, DC, for protesting segregation in South Africa
2006	Dies on January 30 at age seventy-eight